Fighter Basing Options to Improve Access to Advanced Training Ranges

BRADLEY DEBLOIS, PATRICK MILLS, ANU NARAYANAN, DARA GOLD,
ANNA JEAN WIRTH, JOHN A. AUSINK, BART E. BENNETT,
RONALD G. MCGARVEY, RAHIM ALI, JAMES SYME, RUSSELL H. WILLIAMS

Prepared for the Department of the Air Force
Approved for public release; distribution unlimited

RAND PROJECT AIR FORCE

For more information on this publication, visit www.rand.org/t/RRA169-1

Library of Congress Cataloging-in-Publication Data is available for this publication.
ISBN: 978-1-9774-0643-9

Published by the RAND Corporation, Santa Monica, Calif.

Cover: U.S. Air Force photo by Tech. Sgt. Jerilyn Quintanilla.

Preface

The 2018 National Defense Strategy emphasizes the need for the United States to restore warfighting readiness and field a lethal force capable of defeating aggression by a major power. One element in meeting this goal is enabling units to train in an environment that is sufficiently representative of the threats posed by a major power. For U.S. Air Force (USAF) fighter pilots, this means training at ranges with appropriate airspace, threat emitters, targets, and electronic support measures. The USAF has determined that few, if any, existing training ranges have the capabilities to provide fighter pilots with advanced training.

The Office of the Director of Training and Readiness, Deputy Chief of Staff for Operations, Headquarters U.S. Air Force is developing an operational training infrastructure investment plan to upgrade certain ranges with sufficient capabilities to provide fighter pilots with advanced training. This investment strategy shifts the current distributed investment across 30-plus ranges to a more-focused investment in fewer ranges. In addition to range upgrades, the USAF may also consider potential fighter squadron restationing options that would improve access to upgraded training ranges.

This report details RAND Project AIR FORCE's (PAF) framework, tools, and analysis showing the potential effectiveness of different combinations of range upgrades and squadron restationing while also considering costs and risks. The research reported here was commissioned by the Principal Deputy Assistant Secretary of the Air Force for Installations, Environment, and Energy and conducted within the Resource Management Program of PAF as part of a fiscal year 2019 project entitled *Optimal Basing Posture for U.S.-Based Forces*.

RAND Project AIR FORCE

RAND Project AIR FORCE (PAF), a division of the RAND Corporation, is the Department of the Air Force's (DAF's) federally funded research and development center for studies and analyses, supporting both the United States Air Force and the United States Space Force. PAF provides the DAF with independent analyses of policy alternatives affecting the development, employment, combat readiness, and support of current and future air, space, and cyber forces. Research is conducted in four programs: Strategy and Doctrine; Force Modernization and Employment; Manpower, Personnel, and Training; and Resource Management. The research reported here was prepared under contract FA7014-16-D-1000.

Additional information about PAF is available on our website:

www.rand.org/paf/

This report documents work originally shared with DAF on September 25, 2019. The draft report, issued on September 27, 2019, was reviewed by formal peer reviewers and DAF subject-matter experts.

Contents

Figures

Tables

Summary

Issue

To support the goals outlined in the 2018 National Defense Strategy, U.S. Air Force (USAF) fighter pilots need access to live training ranges with airspace, threat emitters, targets, and electronic support measures that are sufficiently representative of the capabilities of potential major power adversaries. The USAF is developing a modernization plan to upgrade training ranges. It might also consider restationing fighter squadrons to improve access to the upgraded training ranges. The USAF asked RAND Project AIR FORCE (PAF) to evaluate range upgrade and squadron restationing policies that could maximize access to advanced training ranges.

Approach

PAF researchers analyzed these policy options through three lenses: effectiveness, cost, and risk. The PAF team developed an optimization model to explore the effectiveness achieved through combinations of range upgrade and squadron restationing policies. A cost assessment provides a comparison of the up-front costs (but not full life cycle costs) associated with range upgrades and squadron restationing. The collection of hazard exposure maps, climate data, and electric power reliability provides a basis for comparing risks to bases and ranges in different parts of the United States.

Conclusions

- Range upgrades alone can provide only a portion of fighter squadrons with access to advanced training ranges. Restationing could significantly increase access, but the amount would depend on institutional freedom to make restationing decisions. Most significantly, if Air National Guard squadrons cannot be consolidated near advanced training ranges, the potential benefits of restationing would be substantially limited.
- Using the current basing posture and planned range upgrades, the F-22 squadrons may not have access to advanced training ranges.
- The largest opportunity to improve readiness in the long term is integrating the range modernization plan and the F-35 rollout.
- The one-time cost for restationing a fighter squadron and the cost to procure equipment for a single range modernization are on the same order of magnitude. However, when research and development and operation and sustainment costs are taken into account, range upgrades may be substantially more expensive over the long term. Upgrading a single range may provide access for more than one squadron, and a cost-effectiveness

assessment should be conducted that accounts for the life cycle range modernization costs.

- There is significant variability in electric power reliability and exposure to natural hazards and climate effects across USAF fighter bases and ranges that might require different levels of investment to recover from or mitigate disruptions.

Recommendations

The USAF should consider

- prioritizing a range upgrade near an F-22 base and consolidating F-22 squadrons. This would require a more-detailed analysis of airfield capacity issues, range capacity, and availability constraints.
- coordinating the introduction of new F-35 squadrons, retirement of legacy aircraft, and range upgrades to ensure that F-35 squadrons would have range access at the earliest possible time.
- developing a training strategy that outlines how much training would be required at each range capability level to better understand how much range capacity would be required and then evaluate restationing against other potential solutions.
- developing full life cycle cost estimates for range modernization to understand the number of ranges that would be affordable over the long term and how those costs would compare with the cost and institutional challenges of restationing squadrons.
- collecting and incorporating relevant risk data, such as hazard exposure maps, climate data, and electric power reliability metrics, in basing decisions.

Acknowledgments

We thank Richard Hartley, Principal Deputy Assistant Secretary of the Air Force for Installations, Environment, and Energy, for sponsoring this work. We are thankful to Hartley and his staff for their guidance and critical review of our work throughout the project. Principal contributors from Hartley's office include Col Scott Bryant, Col Craig Rezac, Terry Tallent, Tim Brennan, Jim Sample, Ed McCarthy, and Bill Spacy. We also benefited greatly from frequent interaction and input from leadership and staff from the Office of the Director of Training and Readiness, Deputy Chief of Staff for Operations, Headquarters U.S. Air Force, including Maj Gen James Jacobson (and his predecessor, Maj Gen Scott Smith), William MacLure, Edward Chupein, and Andy Chamblee. We are also grateful for the force structure data provided by John Butcher at AF/A5/8.

Air Combat Command (ACC) and Air Force Installation and Mission Support Center (AFIMSC) made valuable contributions to the project. Although the contributors from ACC are too numerous to name, we would like to extend a special thanks to Ted Uchida of ACC, Directorate of Air and Space Operations (ACC/A3), Sam Nelson (ACC/A3AR), Greg Calhoun (ACC/A3TO), and Col Uduak Udoaka (ACC/A4R). Primary contributions from AFIMSC include Peter Senior, Gregory Peterson, Derrick Hood, and Gordon Robertson.

Within the RAND Corporation, the authors are thankful to Obaid Younossi and Jim Powers for their program leadership and strategic guidance. Mike Lostumbo, Paul Dreyer, and Tony DeCicco conducted a review of the technical methodology that improved the quality of the analysis. Bill Taylor and Jeff Kendall reviewed an earlier draft of this report that substantially improved the quality of the report. Jim Leftwich and John Drew provided substantial contributions as senior advisers to the project.

Acknowledgment of these individuals does not imply their endorsement of the views expressed in this report.

Abbreviations

ACC	Air Combat Command
ACC/A3	Air Combat Command, Directorate of Air and Space Operations
AF/A3T	Office of the Director of Training and Readiness, Deputy Chief of Staff for Operations, Headquarters U.S. Air Force
AFB	Air Force Base
AFI	Air Force Instruction
AFIMSC	Air Force Installation and Mission Support Center
AFMAN	Air Force Manual
AFSBP	Air Force Strategic Basing Process
AFWIC	Air Force Warfighting Integration Capability
ANG	Air National Guard
BMC	basic mission capable
BMGR	Barry M. Goldwater Range
BOS	base operating support
BRAC	Base Realignment and Closure
CATCODE	category code
CMR	combat mission ready
CONUS	continental United States
CT	continuation training
DOC	Designed Operational Capability
DoD	U.S. Department of Defense
EIA	Energy Information Administration
FOR	Forced Outage Rate
FTU	formal training unit

FY	fiscal year
GCM	General Circulation Model
JBER	Joint Base Elmendorf-Richardson
JPARC	Joint Pacific Alaska Range Complex
LVC	live-virtual-constructive
MACA	Multivariate Adaptive Constructed Analogs
MDS	mission design series
NERC	North American Electric Reliability Corporation
NOAA	National Oceanic and Atmospheric Administration
NTTR	Nevada Test and Training Range
OTI	operational training infrastructure
PAA	primary aircraft authorized
PAF	Project AIR FORCE
RAP	Ready Aircrew Program
RCP	Representative Concentration Pathways
RDT&E	Research, Development, Test, & Evaluation
RegAF	Regular Air Force
RTM	RAP Tasking Memoranda
SAIDI	System Average Interruption Duration Index
SAIFI	System Average Interruption Frequency Index
SLR	sea level rise
SME	subject-matter expert
UFC	unit facilities criteria
USAF	U.S. Air Force
USAFR	U.S. Air Force Reserve
USFS	U.S. Forest Service
USGS	U.S. Geological Survey
UTTR	Utah Test and Training Range

1. Introduction

The U.S. Air Force (USAF) manages changes to its basing portfolio through the Air Force Strategic Basing Process (AFSBP), which seeks to provide a transparent, repeatable, and defendable process for making basing decisions (Air Force Instruction [AFI] 10-503, 2017). A previous RAND Project AIR FORCE (PAF) analysis found that the process and data quality underlying the AFSBP largely accomplish these goals (Samaras et al., 2016). However, that same analysis noted that although individual basing decisions include enterprisewide considerations, there is a lack of strategic, portfolio-level thinking and analysis in defining a continental U.S. (CONUS) basing posture.

A subsequent PAF project outlined a framework for making enterprisewide assessments of domestic basing decisions in the context of the CONUS-based F-35A fleet by comparing effectiveness, cost, and risk for alternative basing postures (Bednarz et al., 2016). The objective in that analysis, which was conducted in fiscal year (FY) 2014, was to evaluate postures that minimize life cycle costs at acceptable levels of risk while maintaining readiness. A third PAF project, conducted in FY 2015, extended that work by developing a model to identify basing locations for the F-35A fleet and training locations that minimize enterprisewide flying costs associated with participation in large-scale composite force training exercises (Narayanan et al., 2016). The focus on cost minimization in both studies was appropriate given the post-sequestration fiscal environment at the time that emphasized cost savings and the security environment that balanced ongoing operations in the Middle East with a rebalance of resources to the Pacific. In addition, neither the F-35 training concept of operations nor the requirements for fifth-generation training ranges were yet defined. Changes in U.S. national security policy objectives and developments in fifth-generation training requirements warrant an update to examining domestic basing postures in terms of effectiveness, cost, and risk.

The 2018 National Defense Strategy emphasizes the need for the United States to restore warfighting readiness and field a lethal force capable of defeating aggression by a major power. One element in meeting this goal is enabling units to train in an environment that is sufficiently representative of the threats posed by a major power. For USAF fighter pilots, this means training at ranges with appropriate airspace, threat emitters, ground and air targets, and electronic support measures. Flying in these environments allows pilots to train in conditions consistent with those expected in an actual conflict and to experience firsthand the latest technology threat systems' capabilities. This type of advanced training better prepares pilots to meet operational plan execution requirements and more effectively enables the USAF to organize, train, and equip its forces. The USAF has determined that few, if any, of its existing training ranges have the capabilities to provide fighter pilots with this advanced training.

The Office of the Director of Training and Readiness, Deputy Chief of Staff for Operations, Headquarters U.S. Air Force (AF/A3T) is developing an operational training infrastructure (OTI) investment plan to upgrade certain ranges with sufficient capabilities to provide fighter pilots with a training environment that adequately represents the challenges posed by major powers. This investment strategy shifts the current distributed investment across 30-plus ranges to a more-focused investment in fewer ranges. Fighter squadrons at bases near these upgraded training ranges would then be able to conduct advanced training daily. However, depending on the number of ranges upgraded, there may be some fighter squadrons that are located too far from any of the upgraded ranges to train daily.

In FY 2018, the Office of the Secretary of Defense for Energy, Installations, and Environment drafted an action memo directing the services to analyze force restationing actions that would better align forces and infrastructure to increase readiness and enable the National Defense Strategy. Recognizing that the OTI investment plan may not include range upgrades such that all fighter squadrons would have local access, Principal Deputy Assistant Secretary of the Air Force for Installations, Environment, and Energy asked PAF researchers to analyze potential squadron restationing options that improve the access of fighter squadrons to the upgraded training ranges. This report details the PAF team's framework, tools, and analysis and presents the potential effectiveness of different combinations of range upgrades and squadron moves while also considering costs and risks to inform the range modernization plan and potential restationing decisions.

Project Objective and Approach

The objective of the project is to assess options for improving operational effectiveness of USAF fighter forces by providing greater access to training ranges equipped with replications of advanced threat systems.[1] Access to training ranges is defined, for the purposes of this report, as fighters based within approximately 150 nautical miles (nm) of a range—or roughly the ability to fly to the range, spend at least 45 minutes on the range, and return to the base without being required to refuel. This type of access would allow daily range use for completion of training requirements. Greater access may be achieved by upgrading ranges, moving squadrons to bases closer to upgraded ranges, or a combination of range upgrades and squadron moves. In this report, we explore squadron restationing and range upgrade options that can maximize access while evaluating cost and risk measures associated with these options.

The overall approach comprises an effectiveness, cost, and risk assessment. The primary focus of the project was on the effectiveness analysis, which involved development of an

[1] Details of the planned threat systems are not available to the general public. In general, these systems would be representative of the surface-to-air missile system threats, jammers, and other integrated air defense systems employed by potential major power adversaries.

optimization model that maximizes squadron access to upgraded ranges given an allowable number of squadron movements and range upgrades. The model can either maximize effectiveness via optimal squadron movements given a predetermined set of range upgrades or maximize effectiveness by selecting both range upgrades and optimal squadron movements.

If the set of range upgrades is predetermined, the model accepts the range upgrades as inputs, an existing or future force structure (i.e., the location of existing or future fighter squadrons at current or planned bases), a set of constraints that define allowable moves (e.g., how many additional squadrons can be added to a base, how many mission design series [MDS] types can colocate on a base, can active-duty and air reserve component units colocate), and the number of allowable squadron movements. The model provides as output the squadron movements that maximize effectiveness. In the case where the model generates recommended range upgrades as outputs, simply the number of desired range upgrades is provided in addition to the other inputs. The analysis examined the level of effectiveness achieved through various combinations of range upgrade and squadron restationing policies.

It was outside the scope of this study to complete a full life cycle cost estimate for squadron restationing or range upgrades. However, to lay the foundation for a potential future cost-effectiveness analysis, we developed the up-front, one-time cost estimate for relocating squadrons that includes the infrastructure costs associated with adding a squadron of aircraft to a base (e.g., ramp space, hangar space), infrastructure costs associated with adding people to a base (e.g., dining facilities, family centers), and personnel movement costs.[2] We then compared this estimate with the range upgrade cost estimates provided by AF/A3T to provide some insight into the trade-offs between range upgrades and squadron restationing.

Risk, a third consideration in these restationing decisions, is also the most difficult to take into consideration. Natural hazards, other climate effects, and power disruptions can impede military operations and have significant financial consequences for the USAF. Although it is difficult to predict when or how often these events will occur, their impact can be significant. Planning for these risks using available information could reduce USAF mission impact, spending on post-disaster rebuilding, or both. We gathered publicly available data and information, including hazard exposure maps, downscaled climate projections, and historical electric grid reliability data, that can provide high-level insights into the relative susceptibility of different USAF bases and ranges to different types of hazards and threats.

Scope

The framework and analysis in this report focus on maximizing access to advanced live training ranges for daily use, through additional range upgrades, squadron restationing, or both.

[2] This accounted for adding a squadron of aircraft assigned to the base. It did not account for any additional capacity that would be required to accommodate permanent or visiting adversary aircraft.

The focus of the study is not on the specific OTI upgrades planned for each range or the limitations of training at live ranges, which include insufficient adversary air, inadequate airspace to allow training to full fifth-generation capabilities, and communication difficulties between fourth- and fifth-generation capabilities, because these have been well-documented in previous research.[3] Instead, this project seeks to understand the potential benefits of restationing fighter squadrons, assuming the range upgrades in the OTI investment plan successfully address these limitations.

In addition, it is reasonable to expect that the USAF will need to develop a graduated training strategy in which lower-level training ranges are used to satisfy basic training requirements while highly capable ranges are used for advanced requirements. When such a training strategy develops, the importance of proximity to a range, the primary effectiveness metric in this study, could change. For example, if required training at advanced ranges could be accomplished in a few weeks per year, the USAF could consider temporary deployments or the use of tankers to provide range access. Similarly, advancements in integrated live-virtual-constructive (LVC) capability may allow more training to be done in simulators, therefore reducing the requirement for live range access. The potential of LVC, and the challenges in realizing such capabilities, have also been documented in previous research (Ausink et al., 2011; Ausink et al., 2018). As developments in these areas progress, the effectiveness of restationing needs to be reassessed.

This analysis maintained the basic organization of the U.S. fighter force in terms of number of squadrons and distribution across the active-duty, reserve, and Air National Guard (ANG) bases. Another approach for increasing access to advanced ranges could be to change the number of authorized aircraft in a squadron or the active-reserve mix. However, these alternatives were not analyzed in this analysis. Such changes would also have impacts on pilot absorption and maintenance and support costs (McGarvey et al., 2013).

Finally, although we have used actual or planned basing locations in this analysis, and, in some cases, highlight basing decisions resulting from the modeling framework, we caution the reader to view those as potentially attractive basing actions but not necessarily recommended actions. First, as we will highlight in Chapter 2, desirable basing actions depend significantly on both policy choices and assumptions. In addition, we have not conducted detailed feasibility analysis for any specific basing action. Rather, these analyses and recommendations should inform readers on the enterprise, portfolio-level posture shifts that better align fighter-basing with range upgrades to maximize range access.

Organization of This Report

The remainder of this report is structured as follows:

[3] For more information, see Ausink et al., 2011; Ausink et al., 2018; and Rosello et al., 2019.

- Chapter 2 presents the effectiveness analysis, showing how various combinations of range upgrades and squadron restationing policies interact to achieve certain levels of effectiveness.
- Chapter 3 presents cost analysis, comparing squadron restationing with range upgrade costs.
- Chapter 4 presents a framework for incorporating risk into basing decisions, including examples of pertinent metrics.
- Chapter 5 presents our conclusions and recommendations.

2. Effectiveness Analysis

An effectiveness analysis was conducted to evaluate restationing and range upgrade policies that enable improved access to advanced live training ranges. In this chapter, we describe our methodology, including the definition of effectiveness in the context of this study, the development of an optimization model, and results of the analysis.

Methodology

Defining a Measure of Effectiveness

In this analysis, we measure the effectiveness of a set of range upgrades and an associated fighter basing posture by the number of fighter squadrons that have access to upgraded training ranges.[4] Access to training ranges is defined, for the purposes of this report, as being located within 150 nm of the range—or roughly close enough to fly to the range, spend at least 45 minutes on the range, and return to the base without being required to refuel.[5] This is a planning factor used by AF/A3T in analysis that defines *access* without incorporating details associated with each MDS type,[6] allowable air routes and altitudes, range entry points, and other factors.

Our effectiveness calculation is slightly more nuanced than simply counting the number of fighter squadrons within 150 nm of upgraded ranges. During discussions with subject-matter experts (SMEs) at AF/A3T and Air Combat Command, Directorate of Air and Space Operations (ACC/A3), it became clear that it was not equally necessary for different MDS types to have access to upgraded ranges. For example, it may be more valuable for an F-35 squadron to have access to an upgraded range, given its capabilities and intended missions, than an F-15C squadron. In addition, it may be more valuable for an operational squadron to have access to an upgraded range than a training squadron. It is important to capture these differences in any

---

[4] In our analysis, effectiveness is calculated at the squadron level. We do not scale effectiveness by the primary aircraft authorized (PAA). We also calculate effectiveness using fighter access to ranges only. We assume that the longer range of other platforms could enable access to the upgraded ranges determined based on fighter-specific analysis, but this should be considered in future analysis.

[5] For example, an F-35A has fuel capacity of 18,000 pounds (lbs), assuming a 20-percent fuel reserve provides 14,400 lbs of useable fuel. Assuming an average fuel burn rate of 10,000 lbs per hour and a cruise speed of 500 knots, 6,000 lbs is used in transit to a range that is 150 nm away, leaving about 8,000 lbs for use on the range (enough for approximately 45 minutes of training, depending on the fuel burn associated with training mission profiles). The useable range time would vary somewhat depending on the specifics for each MDS, but it is the judgment of AF/A3T that being within 150 nm enables all MDS to access the range with sufficient fuel to conduct meaningful training.

[6] *Mission design series* is defined as the "official designation for aerospace vehicles used to represent a specific category of aerospace vehicles for operations, support, and documentation purposes" (AFI 16-401, 2014, p. 16).

measure of effectiveness. The following section describes how we used existing training requirements and range assignments to prioritize access to training ranges based on MDS.

Prioritizing Fighter Aircraft Range Access

USAF fighter pilots maintain either combat mission ready (CMR) or basic mission capable (BMC) flying status while in their flying units. A pilot with CMR status is qualified and proficient in all of the primary missions defined in a unit's Designed Operational Capability (DOC) Statement.[7] BMC status means that a pilot is familiar with, and may be qualified and proficient in some of, the primary missions of the unit.[8] Pilots whose primary duty is to fly maintain CMR status; BMC designations are assigned to pilots who have a primary job performing wing supervision or staff functions that directly support the flying operation.[9] Continuation training (CT) is the training required "to maintain proficiency and improve pilot capabilities to perform unit missions," and the CT requirements for a fighter pilot to maintain CMR or BMC status are described in Ready Aircrew Program (RAP) Tasking Memoranda, or RTMs (AFI 11-2F-16, 2015, p. 50).

The RTM sets the minimum required annual mix of aircraft sorties, simulator missions, and training events that aircrew members must accomplish to maintain combat mission readiness (Air Combat Command, Flight Operations Division, 2018). For example, an inexperienced F-22 pilot must fly 108 aircraft sorties each year (nine per month) and perform 36 missions in a simulator (three per month) to maintain CMR status.

Each aircraft RTM includes a table of flight mission and sortie requirements that describe how those sorties are to be distributed among primary, secondary, and basic skills missions. For example, an inexperienced Regular Air Force (RegAF) F-22 pilot is expected to fly 13 defensive counterair missions (a primary mission), 11 air interdiction missions (a secondary mission), and eight basic fighter maneuver missions (a basic skill) annually to maintain CMR status.

In addition to requirements by mission type, the RTM includes a table that describes the types and numbers of flight events that must be accomplished each year. For example, an experienced RegAF F-22 pilot is expected to accomplish 12 four-ship events and 15 night air-to-air events annually to maintain CMR status. For an F-22, there are 27 event types and 148 total events to be accomplished.

A separate USAF publication, Air Force Manual (AFMAN) 13-212, links these flight events to required range capabilities. Attachment 3 of the manual provides training requirements, the

---

[7] The DOC Statement reflects the unit's core mission(s) and the level of capability for which it was organized (AFI 10-201, 2019, para. 2.11.5). A DOC statement "ensure[s] standards of reporting and is meant to assist units and commanders with gathering and reporting readiness data that is included in the Defense Readiness and Reporting System (DRRS)" (AFI 10-201, 2019, para. 1.4).

[8] CMR and BMC are defined in the AFI 11-2XX series of AFIs, which describes flying training requirements. For example, AFI 11-2F-16, 2015, describes the flying requirements for the F-16.

[9] For more details related to BMC and CMR, see AFI 11-2F-16, 2015.

source of the requirements, and the range capabilities required to accomplish a given training requirement. For example, to accomplish the F-22 RAP requirement for a large force exercise, a range that enables the replication of scenario-based threats is needed (AFMAN 13-212, 2018, Table A3.17). In addition, Attachment 4 of the manual indicates ranges that units are expected to use for certain events that require ranges.[10]

As part of its effort to develop long-term plans for improving training range infrastructure, AF/A3T developed a rating system for ranges based on, among other things, the size of the range, the airspace available in the range, and the threat and communications equipment at the range. Using this scheme, AF/A3T rated all existing USAF training ranges. Only the Nevada Test and Training Range (NTTR) scored well enough based on its existing capabilities to be considered an advanced range that is capable of replicating the threat posed by peer adversaries. By examining which training requirements for different aircraft were assigned to the NTTR, we can infer which types of training are more important to conduct at higher-level ranges.

For each MDS, we calculate the percentage of RTM events that would require an advanced range. We can then normalize the percentages into a weighted score that provides some insight into the relative importance of each MDS with respect to its access to a high-level range. Table 2.1 shows the percentage of range events that would require advanced ranges and the corresponding normalized weighted score for each MDS.

Table 2.1. Advanced Range Requirement by MDS

MDS	Percentage of Range Events That Require Advanced Range	Normalized Weight
F-35	50%	1.00
F-22	35%	0.70
F-16	23%	0.46
F-15C	19%	0.38
F-15E	17%	0.34
A-10	16%	0.32

These weights will be used in the effectiveness analysis to prioritize squadron restationing. Here is one way to interpret the normalized weights in the third column:

- Giving F-35s access to high-capability ranges provides the most effectiveness.
- Giving F-22s access to high-capability ranges provides 70 percent of that effectiveness.
- Because the normalized weight for an F-35 squadron is about double that of an F-16 squadron, it is roughly equally effective for one F-35 squadron to have access or two F-16 squadrons.

[10] For example, the 335th F-15E squadron (Seymour Johnson Air Force Base [AFB] in North Carolina) is assigned the Dare County range for six events (such as chaff, flare, and strafe) but is assigned the NTTR for flag exercises.

- Because the normalized weight for an F-35 squadron is about triple that of F-15C, F-15E, and A-10 squadrons, it is roughly equally effective for one F-35 squadron to have access or three total squadrons of F-15C and/or F-15E and/or A-10.

As mentioned previously, in addition to differences between MDS types, there are differences in the value of access to upgraded training ranges between operational and training squadrons.[11] We rely on military judgment to assess the relative importance of operational versus training squadron access to upgraded training ranges. In particular, SMEs at AF/A3T estimate the value of training squadron access to upgraded ranges to be approximately one-fourth that of an operational squadron. Thus, the weighted score for training squadrons would be one-fourth of the normalized MDS weight shown in Table 2.1.[12]

Given these normalized weights, we can now calculate the measure of effectiveness for any posture as the weighted sum of squadrons within 150 nm of an upgraded training range where weights are derived from MDS-specific range requirements for operational and training squadrons as shown in Table 2.1.

Developing Postures That Maximize Effectiveness Using an Optimization Model

Using our definition of effectiveness, we can see that upgrading a range will increase a posture's total weighted effectiveness score. Additionally, the score increases if

- a squadron that is not within 150 nm of an upgraded range is moved to a base that is within 150 nm.
- a lower-weighted MDS (e.g., an A-10) is swapped with a higher-weighted MDS (e.g., an F-35).

We are interested in which combination of range upgrades and squadron movements provides the highest effectiveness score. Given the number of combinations of range upgrades and squadron movements, this problem lends itself to an optimization model.

The structure of the optimization model to be used here depends upon the set of decisions under the model's purview. If the set of range upgrades is determined in advance (outside the scope of the model's decisionmaking), the model maximizes effectiveness by selecting the optimal basing location for each squadron, subject to a set of constraints limiting allowable basing assignments (the constraints assumed in our analysis will be discussed later). Alternatively, the model could provide recommendations on both which ranges to upgrade and which squadrons to assign to each base, subject to both the aforementioned constraints on squadron basing assignments and additional constraints on allowable range upgrade decisions. It

---

[11] The Air Force also has fighter test and evaluation squadrons. Discussions with SMEs indicated that specific instrumentation of ranges is more important to support these squadrons than threat replication systems in the OTI plan. Thus, test squadrons were assigned to their current base and range and not part of the effectiveness calculation.

[12] Using input from the sponsor, we assumed that training squadrons would not be relocated, so these were locked at their current locations but included in the effectiveness calculation.

should be apparent that greater effectiveness values could potentially be achieved if the model is allowed to optimize both squadron basing assignments and range upgrade decisions in an integrated fashion.[13]

Whether or not the model selects the set of range upgrades, the model's required input data would include

- the sets of squadrons, bases, and ranges under consideration
- the default beddown posture (i.e., the current basing assignment for each squadron under consideration)
- the set of allowed base assignments for each squadron[14]
- the distance between each base-range pair
- the capacity at each base (in terms of the maximum number of squadrons allowed to be assigned)
- the maximum number of squadron movements that are allowed.[15]

If the model is to identify which ranges to upgrade, an additional input data parameter would be needed to specify the number of range upgrades to be performed.

Three types of decision variables can potentially be used, depending on the scenario being optimized. In every case, there is a binary assignment variable indicating a yes/no decision: *Is squadron assigned to base **b** and range **r**?* If the model selects the optimal range upgrades, then another binary variable is used to indicate, *Is range **r** upgraded?* Finally, in the event that the USAF wanted to identify a beddown posture that did not mix different MDS types at a single base but did allow the currently assigned MDS at a base to potentially change, a third binary assignment variable would be used to indicate, *Is MDS **m** assigned to base **b**?*

The model maximizes the effectiveness score, subject to the following constraints on what constitutes a feasible beddown posture:

- Each squadron is assigned to exactly one base and one upgraded range.
- The total number of squadrons assigned to a base cannot exceed the capacity of that base.
- The total number of squadron movements cannot exceed the maximum number allowed.

---

[13] This is necessarily true, unless the set of predetermined range upgrades happened to be an optimal set of range upgrades for a particular instance of this optimization problem.

[14] For example, in a particular problem, one might envision evaluating such policies as "U.S. Air Force Reserve (USAFR) squadrons can only be assigned to USAFR bases" or "F-16 squadrons can only be assigned to current F-16 bases."

[15] The model was also designed to include range capacity, in terms of number of assigned squadrons, as a constraint. However, in our analysis we did not enforce a range capacity constraint. During discussions with SMEs, it became clear that the number of squadrons that could be assigned to a range would depend significantly on the portion of the training that required ground range access, which varies by MDS. More-detailed analysis of training requirements would be required to set meaningful range capacities, which was outside the scope of this project.

Additionally, if the model is to select the set of range upgrades, a constraint is added ensuring that the total number of range upgrades does not exceed its maximum-allowed level. Finally, if it is assumed that a feasible beddown posture cannot mix different MDS at a single base, two constraints are needed to enforce this constraint.

During the development of this optimization model, initial computational testing determined that there were often many different beddown postures that were capable of achieving the maximum-possible effectiveness score. To help select between these alternative optimal solutions, an additional optimization stage was added. Let z denote the maximum-possible effectiveness score for a particular problem instance. This second-stage optimization model would be nearly identical to the first optimization, the only changes being that a new constraint would be added forcing the effectiveness score to be equal to z, and the objective would be changed to minimize the sum of distances between each squadron's assigned base and its assigned range.

Results

Policy Levers and Cases Analyzed

The effectiveness analysis used the methodology just described to explore the effect of two policy levers: range upgrades and squadron restationing. Broadly, we seek to understand how these policy levers interact to achieve various levels of effectiveness. Each broad policy lever includes a set of policy options that will have a different impact on effectiveness.

The range upgrade policy lever contains two categories of options: the number of ranges to upgrade and the priority ordering of the upgrades. In general, we would expect that both of these categories will cause effectiveness to vary, with or without squadron restationing. In this analysis, we consider between one and 17 range upgrades. The upper bound of 17 is based on the number of ranges in the current AF/A3T range modernization plan (AF/A3T, 2019). In addition to the number of ranges upgraded, which ranges are upgraded will affect effectiveness. Ranges that are proximate to several fighter bases may have more of an impact on effectiveness than ranges proximate to just a single base. To show the results from a combination of the number of ranges and which ranges are upgraded, we use "upgrade priority lists." In this analysis, we use a range priority list developed by AF/A3T and a range priority list determined within our modeling framework.

The squadron restationing policy lever contains several interacting options that will affect the potential level of effectiveness achieved. They include

1. the number of squadron moves
2. the total number of squadrons that can be assigned to a base
3. the set of bases under consideration to receive squadrons
4. the number of fighter MDS types allowed per base
5. the rules dictating allowable ANG restationing.

In the first option, we would expect that the more squadrons are permitted to move, the higher the effectiveness score will be. In this analysis, we consider the effectiveness achieved for zero to 20 squadron moves. The number of allowable squadron moves suggests the degree of fighter posture shift that the USAF is willing to consider and support.

In the second option, increasing the total number of squadrons that can be assigned to a base, or the base capacity constraint, will also tend to increase the effectiveness score. In an extreme but impractical case, all squadrons could be relocated to a single base near an upgraded range. Using conversations with SMEs and data from the Air Force Installation and Mission Support Center (AFIMSC), we chose to analyze cases in which active-duty bases could add one or two squadrons, which we will refer to as +1 or +2 over its existing capacity (details of the analysis provided by AFIMSC are provided in Chapter 3).

In the third option, the set of bases that can receive new squadrons will also affect the level of effectiveness. Broadening the set of bases under consideration may increase the number of bases within 150 nm of upgraded ranges and therefore increase the potential benefit of restationing. We chose to analyze a case that includes allowing restationing only to existing fighter bases ("Fighter") and a case that includes existing fighter bases and any active-duty bases that currently have a flying mission ("Fighter plus active nonfighter").

The fourth option analyzed was whether different fighter MDS types can colocate at the same base. We considered three different options: (1) bases can only receive an MDS that already exists at the base (thus, one MDS per base dictated by an existing MDS type, or "1 existing"), (2) bases can have just one MDS but not necessarily of the existing type (thus, an existing MDS could be moved out and a new MDS moved in, or "1 any"), and (3) bases can receive an MDS of different types (thus, multiple MDS types could exist on a base, or "multiple").

Finally, we considered different options for ANG restationing. We examined cases in which the ANG cannot move (referred to as "No move"); ANG units can move, but the same number of units must be at each ANG base before and after (thus, ANG units can "swap" bases); ANG units can move to active-duty bases but not vice versa (referred to as "ANG to Active"); and ANG units can move to active-duty bases and vice versa (referred to as "Free," which indicates that squadrons can move freely between bases).

The policy options examined, and the cases analyzed for each, are summarized in Table 2.2.

Table 2.2. Cases Analyzed for Each Policy Option

Policy Lever	Policy Option	Cases Analyzed
Range upgrades	Number of ranges	1–17
	Range upgrade priority	AF/A3T priority; model-generated priority
Squadron restationing	Number of squadron moves	0–20
	Base capacity constraint	+1, +2 over existing
	Bases considered	Fighter; Fighter plus active nonfighter
	MDS types per base	1 (existing), 1 (any), multiple
	ANG assumptions	No move; Swap; ANG to Active; Free

Force Structure

A key input for our analysis is the starting force structure to analyze. In this analysis, we select a single snapshot in time as defined by the current 2025 force structure projection. The specifics of this force structure are shown in Appendix A. Choosing 2025 allows us to include all of the announced future F-35 basing decisions but is near term enough to determine whether the current fighter force structure could benefit from shifting around upgraded training infrastructure. A clear extension of this work, discussed in Chapter 5, would be to analyze a point further in the future, perhaps to inform basing decisions in the context of the F-35 fielding plan over the next 20 years.

Given the 2025 force structure, as shown in Appendix A, and the set of MDS weights shown in Table 2.1, we can calculate the maximum potential effectiveness for this force structure.[16] Maximum effectiveness, if all squadrons are within 150 nm of an upgraded range, is the weighted sum across all squadrons. For this force posture, the maximum effectiveness score is 31,[17] which we can then normalize to such that we can easily see what fraction of maximum effectiveness various solutions can achieve.

---

[16] Test and evaluation squadrons are part of the fighter force structure that also use training ranges. However, these squadrons rely more on specific instrumentation of ranges than on representative threat systems, so we did not include them in our effectiveness score calculations. However, they did count against the base capacity, and we assumed that they would not move from their current locations.

[17] Each squadron within 150 nm of an advanced range is assigned an effectiveness value based on the weight in Table 2.1, and a score of 0 if not within 150 nm. The absolute maximum effectiveness would be achieved if all squadrons were within 150 nm, so the maximum score of 31 is the sum of the individual effectiveness scores across all fighter squadrons. This is then normalized so that maximum effectiveness is equal to 1.

Findings for AF/A3T Range Upgrade Priority

We begin by examining potential effectiveness of squadron restationing policies given the existing AF/A3T range upgrade priority.[18] The range upgrade priority is shown in Table 2.3.[19]

Table 2.3. AF/A3T Range Upgrade Priority

Upgrade Order	Range	Upgrade Order	Range
1	Nevada Test and Training Range	10	Smoky Hill Range
2	Joint Pacific Alaska Range Complex	11	Air Force Dare County Range
3	Utah Test and Training Range	12	Snyder Electronic Warfare Site
4	Belle Fourche Electronic Scoring Site Range	13	Grand Bay Range
5	Poinsett Range	14	Melrose Range
6	Eglin Test and Training Complex	15	Barry M. Goldwater Range
7	Adirondack Range	16	Warren Grove Range
8	Mountain Home Range Complex	17	Claiborne Range
z	Hardwood Range		

In the following example, we consider a single model run where nine ranges are upgraded, nine squadrons are moved, and the base capacity constraint is one additional squadron, allowing only restationing to existing fighter bases, one MDS per base selected by the model, and no ANG restationing permitted. Figure 2.1 shows the resulting squadron moves that result in maximum effectiveness for this set of policy choices. The colored diamonds are the nine upgraded ranges. The colored squares represent the existing fighter bases that are within 150 nm of each upgraded range. The colored arrows represent the nine squadron moves that maximize effectiveness. A USAFR F-35 squadron was moved from Fort Worth, Texas,[20] to Hill AFB in Utah. The active-duty F-16 squadrons were moved from Shaw AFB in South Carolina to other allowable locations near an upgraded range, and four F-22 squadrons were moved to Shaw. The reasons for these moves are covered in more detail later in this section (see Figure 2.4 and accompanying discussion). Figure 2.1 shows the results for one set of policy options. We explored more than 4,000 combinations of policy options, each providing a result.

---

[18] The AF/A3T upgrade priority was based primarily on providing access to advanced ranges for high-priority squadrons.

[19] We assume that all of the ranges in the AF/A3T upgrade list have sufficient airspace to enable appropriate use of the advanced threat systems. As of the conclusion of this project, AF/A3T and ACC/A3 were in the process of conducting more-detailed assessments of airspace considerations. A possible implication would be that it may not be possible to upgrade a range if there is not appropriate airspace. This could readily be accounted for in the model when that analysis is available.

[20] As of the time of this report, the location of the USAFR F-35 was yet to be determined, but we chose Fort Worth in this analysis for illustrative purposes.

Figure 2.1. Results for a Single Model Run

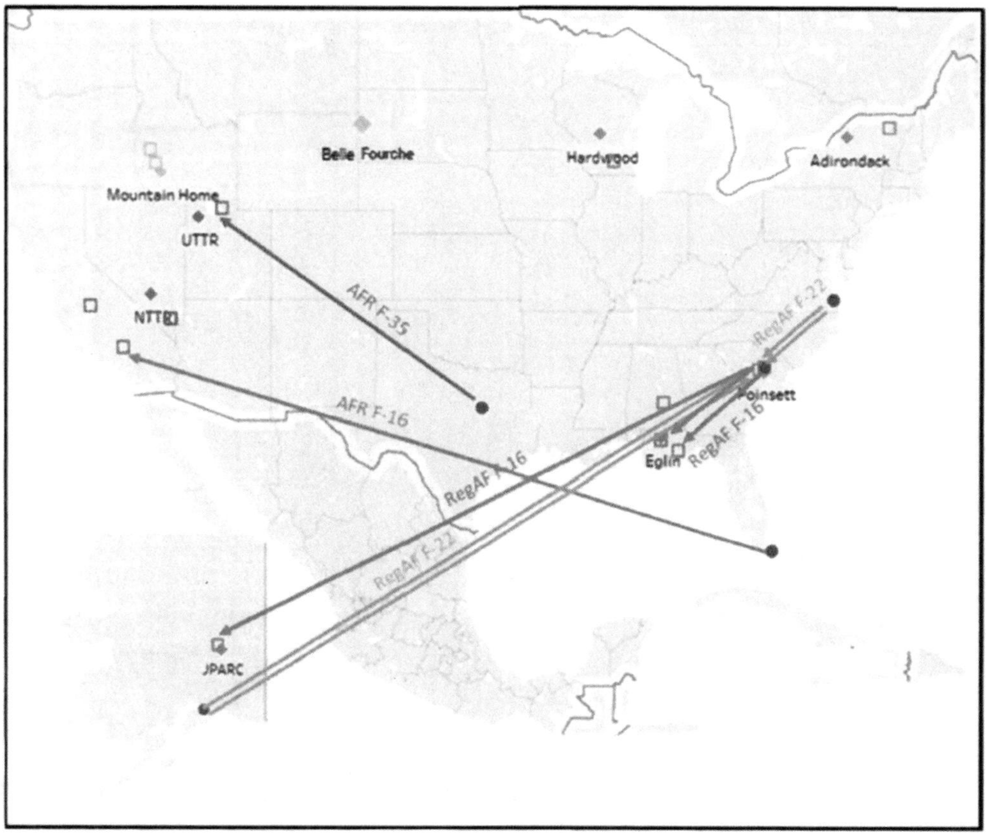

Figure 2.2 shows the effectiveness achieved for the number of range upgrades and number of squadron moves for every combination of restationing policies. Each point on the figure represents the calculated effectiveness (y-axis) for the optimal set of squadron moves given the number of range upgrades (x-axis) and the number of squadron moves (indicated by the color of the dot). As shown in the figure, effectiveness increases as more ranges are upgraded (i.e., moving left to right) and as more squadrons move closer to upgraded ranges (i.e., moving from bottom to top for each number of upgrades).[21] The dark blue circle highlights the example of nine ranges upgraded and nine squadrons moved, as shown in Figure 2.1.

---

[21] As mentioned previously, specific range capacities per MDS were not known at the time of the study, so we did not enforce a range capacity constraint. One implication could be that we are overestimating the effectiveness of certain solutions, particularly those where many squadron moves are used to consolidate around relatively few ranges. As range capacity estimates become available, these should be added into the framework.

Figure 2.2. Effectiveness for Squadron Restationing Policies Given AF/A3T Range Upgrade Priority

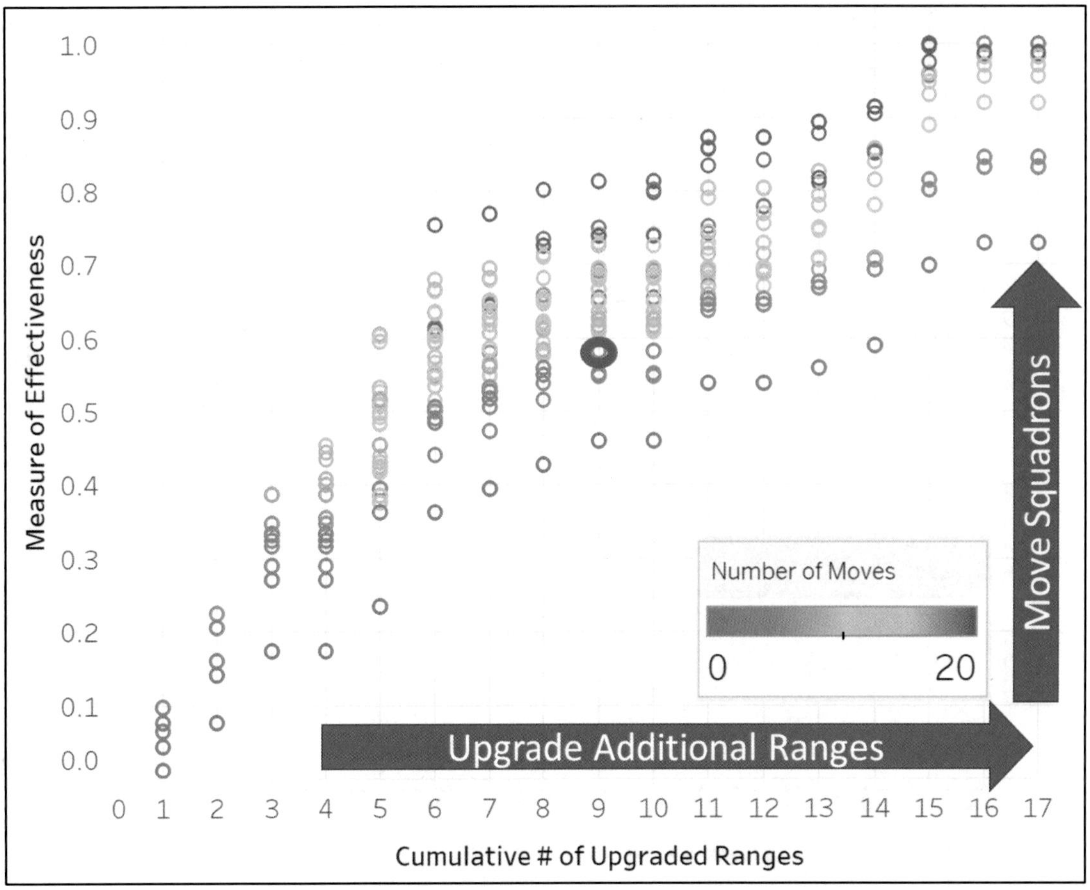

Several findings emerge:

- **Achieving 90-percent–plus effectiveness requires 15 or more of the planned range upgrades and ten to 20 squadron moves.** The only points that exceed 90 percent of max effectiveness appear at the top right of the chart. These solutions include 15 to 17 range upgrades and ten to 20 squadron moves.
- **To achieve 70-percent effectiveness, trade-offs between range upgrades and squadron moves are possible.** Solutions that achieve 70-percent effectiveness include those where 15 to 17 ranges are upgraded and no squadrons move and those where just six to eight ranges are upgraded and 15 to 20 squadrons move. This indicates some trade-off in effectiveness between range upgrades and squadron moves.
- **The largest potential benefit of restationing occurs when roughly half of ranges are upgraded.** If too few ranges are upgraded, there are not enough bases within 150 nm of a range to allow many squadron moves, and therefore the benefit of squadron restationing is limited. Conversely, if all ranges are upgraded, then effectiveness levels are already fairly high, and the additional benefit of squadron restationing is somewhat limited. However, when some middle number of ranges are upgraded, the additional benefit of restationing can be substantial.

The previous results focus on the trade-offs between range upgrades and the number of squadron moves but not the differences between squadron restationing policies. Figure 2.3 shows that restricting ANG squadron restationing limits the options for increasing effectiveness. The left chart in the figure shows potential solutions when ANG movement to active-duty bases is allowed (the difference between the ANG to Active case and the Free case is very small, so we excluded the Free case), and the right chart shows the potential solutions when the ANG is not allowed to move,[22] maintaining the dotted blue line to show the maximum effectiveness when including ANG moves. Given that so much of the fighter force resides in the ANG, if consolidating ANG units around upgraded ranges is not permitted, then there is a fixed upper limit to achievable effectiveness.

Figure 2.3. Restricting ANG Squadron Restationing Limits Options for Increasing Effectiveness

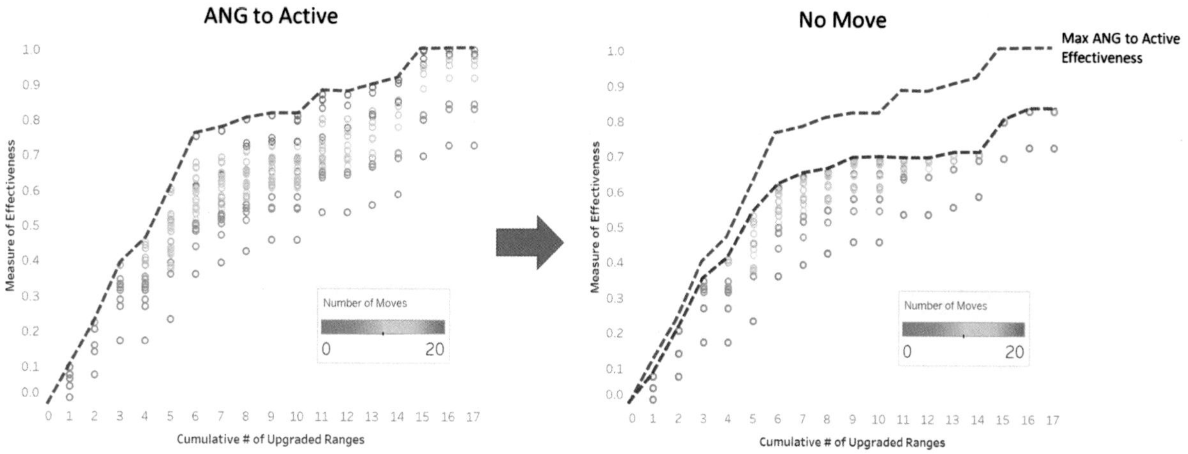

In addition to ANG movement policies, we explored the effects of various policy options, including base capacity, inclusion of nonfighter bases, and allowing multiple MDS types per base. Increasing the number of squadrons that can be added to an existing base increases the potential effectiveness for a given number of range upgrades. The number of squadrons that can be added to a base is likely to be based on a combination of policy considerations (e.g., a wing of fighters is typically not more than four squadrons, so what are the implications if more than four squadrons are located on a single base) and physical constraints (e.g., is there existing excess capacity, or is there developable land to add capacity). A specific capacity assessment for each base was beyond the scope of this study, so we chose to consider allowing one or two additional squadrons per base. The effect of including nonfighter bases was minimal. The existing fighter bases are better aligned with range infrastructure, and the model tended to find solutions within the existing fighter bases. Similarly, we explored the effects of the number of fighter MDS types per base. Limiting to just the existing MDS significantly limits the potential benefit of

---

[22] Also, the difference between the No Move and the Swap cases was very small, so we excluded the Swap case.

restationing. Allowing the model to select a single MDS per base was as effective, in most cases, as allowing multiple MDS per base.

The analysis just presented represents the results of thousands of model runs. As a result, it is difficult to present any single set of recommended squadron movements because those recommendations depend so significantly on how many ranges are upgraded, how many moves are made, move policies, and other unresolved uncertainties (e.g., what is the base-specific capacity constraint). However, in general, the model is seeking to move high-priority squadrons (as dictated by the weighting scheme presented previously) from bases not within 150 nm of an upgraded range to bases within 150 nm of an upgraded range. Using the measure of effectiveness as defined in this study, the preferred move would be to move an F-35 squadron that is not within 150 nm of any of the 17 ranges to a base near one of the highest-priority ranges because that would be advisable under the widest set of assumptions. We can extend that logic to include prioritizing moves for squadrons within 150 nm of only the lowest-priority ranges to bases near the highest-priority ranges. Figure 2.4 shows which squadrons the model chooses to move frequently and the locations that are typically the destination. The specific squadron to base assignment varies widely according to the policy options for a specific case.

Figure 2.4. Frequently Moved Squadrons and Frequent Destinations

Frequently moved squadrons

Comp	Location	Squadron	MDS
Active Duty	JBER	90th FS	F-22A
	JBER	525th FS	F-22A
	Langley	27th FS	F-22A
	Langley	94th FS	F-22A
	Holloman	311th FS	F-16
	Holloman	314th FS	F-16
Reserves	Ft. Worth	OPS-	F-35A
	Homestead	93rd FS	F-16
Guard	Hickam	199th FS	F-22A
	Truax Field	OPS-	F-35A
	Multiple	Multiple	F-16

Frequent destinations

Location	Current MDS
Nellis	Various
Edwards	Various
Eielson	F-35A
Hill	F-35A
Shaw	F-16
Eglin	Various
Tyndall	F-35A
Mountain Home	F-15E

Several observations can be made from this typical model behavior:

- **F-22 squadrons are frequently moved from Joint Base Elmendorf-Richardson (JBER) in Alaska.** The Joint Pacific Alaska Range Complex (JPARC), the second-highest-priority range, is 180 nm away and, therefore, not a viable range given the 150 nm distance constraint used in this analysis. The obvious question is whether a better alternative would be for those squadrons to fly the extra 30 nm to conduct training at JPARC. That is certainly a potential option—although as a result of the longer transit distance, they would spend less time on the range per sortie, so the level of effectiveness may not be equivalent.

- **F-22 squadrons are frequently moved from Joint Base Langley-Eustis in Virginia.** The only air-to-ground range within 150 nm of Joint Base Langley-Eustis is Dare County in North Carolina, currently No. 11 in the AF/A3T range priority. In cases where fewer than 11 ranges are upgraded, the F-22s often move to another location. One obvious question emerges here: Should the range be higher in the upgrade priority? This is a point we revisit in the next section.
- **To accommodate the F-22s, a fourth-generation fighter base near a high-priority range (e.g., Shaw or Mountain Home AFB in Idaho) is converted to an F-22 base.** Given that no existing F-22 base (JBER, Joint Base Langley-Eustis, and Hickam) is near a high-priority range, the model commonly moves the fourth-generation aircraft out of a base like Shaw and replaces them with F-22s. We observed this behavior in the example shown in Figure 2.1.
- **The USAFR and ANG F-35 squadrons are not near high-priority ranges, so they often move to existing F-35 bases.** The F-35 squadron at Fort Worth[23] is not near any of the 17 ranges, so it is moved in essentially all solutions, typically to Eielson AFB in Alaska or Hill AFB in Utah. Truax Field in Wisconsin is near Hardwood Range, currently No. 9 on the priority list—therefore, when fewer than nine ranges are upgraded, Truax is also moved to an existing F-35 base.
- **The remainder of the moves typically involve consolidating ANG F-16 squadrons.** After the F-22 and F-35 moves discussed earlier, the model typically consolidates F-16 ANG squadrons, if ANG restationing is permitted. As F-16 squadrons are replaced by F-35 squadrons going forward, this will become an even more attractive option.

Findings for Model-Generated Range Upgrade Priority

As discussed in the methodology section, the model can either accept a range upgrade priority as input or be used to optimize the combination of range upgrades and squadron moves. When the model is being used to determine the number of ranges and which specific ranges to upgrade, it is possible that the range upgrade priority ranking will vary depending on the number of squadron moves and other restationing policy choices, making it difficult to compare with the AF/A3T upgrade priority. In addition, because decisions on squadron restationing policies are not likely to be fully known prior to the start of range modernization, any upgrade priority that is too specific to a certain set of assumptions may not be especially valuable. Therefore, we ran the model for a variety of squadron restationing policies and generated a single range upgrade priority by aggregating the results such that the most frequently upgraded ranges appear higher on the priority list. This range upgrade priority is not likely to be optimal for any specific set of

---

[23] As of the time of this report, the location of the USAFR F-35 was yet to be determined, but we chose Fort Worth in this analysis for illustrative purposes.

policies but should be more optimal than a set of range upgrades developed without accounting for possible squadron movement.

Table 2.4 shows the model-generated upgraded priority. The first two ranges, the NTTR and the JPARC, are both planned to be key regional and large force training ranges and will certainly receive required upgrades. Therefore, we remove them from the decision space by forcing them to be the top two priorities and allow the model to then select the remaining upgrades. The parenthesis next to each range name shows the change in rank from the original AF/A3T upgrade priority. Figure 2.5 compares the resulting effectiveness scores for each range upgrade priority.

Table 2.4. Model-Generated Range Upgrade Priority

Upgrade Order	Range	Upgrade Order	Range
1	Nevada Test and Training Range	10	Mountain Home Range Complex (-2)
2	Joint Pacific Alaska Range Complex	11	Grand Bay Range (+2)
3	Barry M. Goldwater Range (+12)	12	Warren Grove Range (+4)
4	Eglin Test & Training Complex (+2)	13	Melrose Range (+1)
5	Air Force Dare County Range (+6)	14	Snyder Electronic Warfare Site (-2)
6	Poinsett Range (-1)	15	Belle Fourche Electronic Scoring Site Range (-11)
7	Hardwood Range (Volk Field) (+2)	16	Smoky Hill Range (-6)
8	Adirondack Range (-1)	17	Claiborne Range
9	Utah Test and Training Range (-6)		

Figure 2.5. Comparing AF/A3T and Model-Generated Range Upgrade Priorities

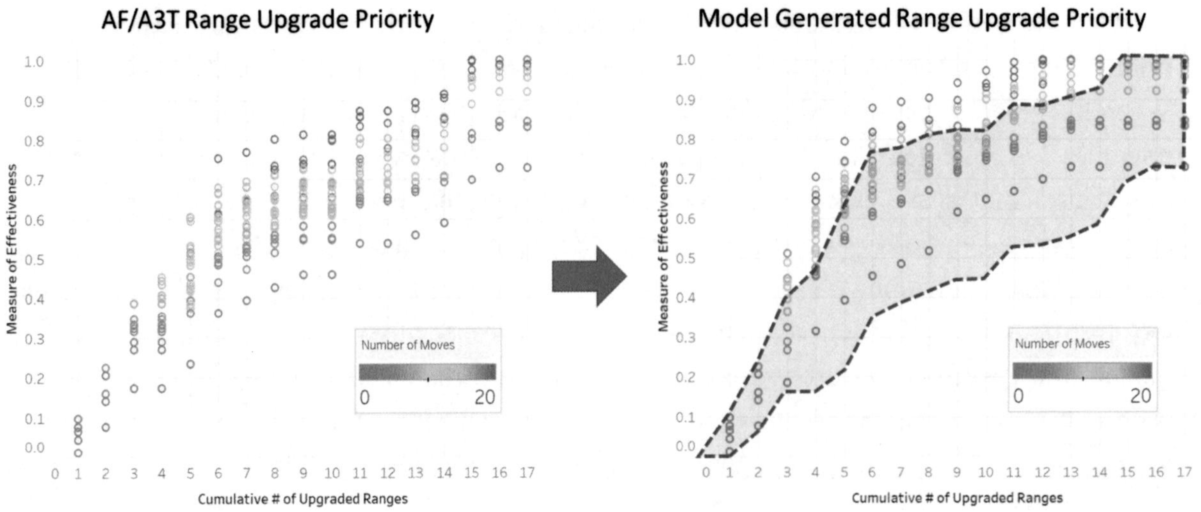

Figure 2.5 shows that the model-generated range priority is more effective; higher levels of effectiveness are possible for the same number of range upgrades. The model-generated range upgrade priority has a few key differences that contribute to this increase in effectiveness:

- **Belle Fourche (South Dakota), Smoky Hill (Kansas), Snyder Electronic Warfare Site (Texas), and Claiborne (Louisiana) are upgraded last.** These four ranges do not have any fighter bases within 150 nm and thus cannot contribute to effectiveness as measured in this analysis. In the model-generated priority, these would be upgraded last, allowing for more-beneficial ranges to be upgraded earlier.
- **Barry M. Goldwater Range (BMGR) in Arizona is upgraded much sooner.** Three bases (Luke, Davis-Monthan, and Tucson) are within 150 nm of BMGR. Luke houses the active-duty formal training unit (FTU) squadrons for the F-16 and F-35. Although training squadrons are given only 25 percent of the value of operational squadrons in our effectiveness metric, the sheer number of training squadrons at Luke (eight) makes it an attractive base to locate an upgraded range nearby. This raises the issue of whether FTU squadrons will actually use the upgraded ranges. If not, the weighting for these units should be adjusted accordingly. BMGR becomes a higher-priority range when adding three A-10 squadrons from Davis-Monthan and two F-16 squadrons from Tucson.
- **Dare County is upgraded much sooner.** Recall from the previous section that, in the AF/A3T priority, there is no F-22 base within 150 nm of a high-priority range. Dare County Range is near enough to Joint Base Langley-Eustis that it becomes an attractive range to upgrade sooner. The model tends to upgrade this range and then add one or two additional F-22 squadrons to Joint Base Langley-Eustis, subject to the policy constraints.
- **The Utah Test and Training Range (UTTR) is upgraded much later.** Given that the UTTR is the only range close to Hill AFB (also in Utah) and its three F-35 squadrons, this initially appears surprising. However, upon closer examination, Hill is the only base within 150 nm of UTTR. So, when potential squadron restationing is included, it becomes more effective for the model to upgrade a range with more nearby bases (e.g., BMGR) and then move the F-35s from Hill to one of those bases (e.g., Davis-Monthan). This highlights how different decisions are made if decisionmakers are jointly considering range upgrades and squadron restationing rather than each in isolation.

As noted previously, this model accounts for just one dimension of effectiveness that drives range upgrade and squadron restationing decisions and fighter access to ranges. The model does not account for all range limitations, such as capacity and airspace. Clearly, there are many other considerations that go into these decisions; for example, Belle Fourche is an important bomber range, so the benefit of upgrading it is not captured in this analysis. The model-generated upgrade priority should not be viewed as a recommended upgrade priority but rather as a means to identify some important range upgrade considerations that arise when specifically focusing on fighter access to ranges and the other assumptions used in this analysis.

3. Cost Analysis

In this chapter, we present the methodology and results for estimating the one-time cost for relocating squadrons, including the infrastructure costs associated with adding aircraft to a base (e.g., ramp space, hangar space), the infrastructure costs associated with adding people to a base (e.g., dining facilities, family centers), and the personnel movement costs. We then compare this estimate with the range upgrade cost estimates provided by AF/A3T to provide some insight into the cost trade-offs between range upgrades and squadron restationing.

Restationing Costs

The one-time restationing cost has two components:

1. *Personnel move costs*, which are the costs to move active-duty unit personnel, their families, and their personal belongings to their new locations.[24]
2. *Infrastructure costs*, which are the costs to expand or upgrade infrastructure at the receiving base to accommodate the new aircraft and personnel.

Relocating squadrons also has long-term operating and sustainment implications, mainly in the form of potential changes to flying profiles as a result of being nearer to or farther from assigned training ranges, and thus sortie duration and flying hour costs. In consultation with the sponsor, we set aside these potential flying costs, with the assumption that, for the foreseeable future, units would maintain the standard training sortie durations and would take any time savings from shorter transit distances in the form of increased productive time on range, and thus increased readiness benefit rather than cost savings.

Personnel Move Costs

Estimating the total cost associated with moving personnel requires an estimate of the total number of people moved and the cost per person moved.

Estimating the Number of Personnel Moved When Restationing a Squadron

Estimating the total number of personnel who would move from one base to another in the event that a flying squadron moves is not entirely straightforward. Most of the personnel who would relocate with a given fighter squadron are not actually "in" the fighter squadron. Thus, for

[24] We excluded the movement of unit equipment from one location to another (e.g., maintenance shops). We assume that this would happen via large-scale ground transportation and would be fairly negligible relative to the other costs we include. Our sponsor confirmed this assumption.

the flying wings in question, we drew from several squadrons on each base to encompass all of the appropriate operations and support activities. These included the following types of units:[25]

- flying squadron
- operations support
- aircraft maintenance
- equipment maintenance
- component maintenance
- munitions maintenance.

We did not include group- or wing-level personnel in this unit list, although some fraction of group personnel would likely shift to the receiving base to provide management oversight.

In this personnel estimate, we focused primarily on Air Combat Command (ACC) units. Another major active-duty group was test units, but their cases were often complicated. Many test bases have more-diverse aircraft types and aircraft numbers than ACC units, making it more difficult to create "clean" calculations for the MDS within our scope. ACC bases, being more homogeneous, made for more-straightforward calculations. The bases we included in our calculation included Davis Monthan, Hill, Holloman (New Mexico), JBER, Joint Base Langley-Eustis, Moody, Mountain Home, and Seymour Johnson. In all, we included 19 fighter squadrons, including the associated support and maintenance units mentioned previously. We used several different approaches to develop average numbers, including regression analysis, scaling to the number of aircraft, and scaling to the number of pilots. Our estimates ranged from 690 to 795 personnel per squadron, with an average of 735 personnel across these squadrons. These numbers are consistent with past research on basing costs.[26]

That estimate included only operations and maintenance personnel. Most installations require a complement of base operating support (BOS) personnel to support the complete base population and infrastructure, much like running a small city.[27] This includes engineering, medical, and numerous administrative functions. Past research found a relationship between operations personnel (i.e., those in flying and other operational units) and support personnel (i.e., BOS functions devoted to running the base itself), such that every operations position "carried" with it 0.4 support positions.[28] Thus, we use the 0.4 multiplier to arrive at a total number of personnel to whom to apply move costs. With our average of 735 operations and maintenance personnel and 294 support personnel (735 * 0.4), our total is 1,029.

---

[25] Manpower and personnel data obtained by PAF in 2019.

[26] Mills et al., 2013, used an F-16 squadron as a case, and estimated 668 personnel with similar organizational units. F-16s are the smallest of the fighter airframes and have only one engine, so their squadron sizes are on the low end of the range for the various MDS we analyzed in the current report.

[27] One exception to this, technically, is many ANG and reserve units. Many are colocated at airports and rely on the civilian and commercial infrastructure, where active-duty bases and units provide the full complement of base services organically.

[28] See, for example, Mills et al., 2013, and Lostumbo et al., 2013.

Estimating the Cost to Move Personnel

Next, we determine the cost to move personnel. To estimate the cost of relocating the personnel in each flying squadron and associated maintenance and support squadrons, we drew data from the USAF's FY 2019 Budget Estimates for Military Personnel Appropriation (USAF, 2018). We used the budgeted funding for permanent change of station moves in 2019. Table 3.1 captures these figures. In each row, we divide the cost of all moves in that category by the number of moves to calculate the average cost per move.

Table 3.1. Summarized Costs of USAF Personnel Moves

Category	Number of Moves	Cost of All Moves ($M)	Cost per Move ($)
Officer	7,400	126.692	17,121
Enlisted	12,000	147.631	12,303

SOURCE: USAF, 2018.

Estimating Per-Squadron Move Costs

We can then estimate the total per-squadron move cost by multiplying the number of personnel moved by the cost per move. In the unit types considered (flying squadron plus associated support and maintenance units), 94 percent of personnel are enlisted.[29] Thus, the average cost per move is a weighted average and results in $12,574 per move. Multiplying that by our constructed squadron, we arrive at a per-squadron move cost of about $13 million.

Infrastructure Costs

Restationing squadrons would also incur infrastructure costs, which consist of the expansion or upgrading of infrastructure at the receiving base to accommodate the new aircraft and personnel. Figure 3.1 shows the various elements of infrastructure cost for squadron moves and our approaches to estimating them. The flow starts at the top, where a squadron being added to a receiving location ("additional squadron") brings with it aircraft and personnel. We assume 24 aircraft per squadron across each fighter MDS. We derived the number of personnel who come with a squadron by integrating several data and information sources, as discussed in the previous section.

[29] Analysis of FY 2018 manpower data indicated that 94 percent of active-duty ACC personnel authorizations in relevant categories (operations and maintenance) are enlisted. This is slightly higher than the 85 percent enlisted across all ACC and 81 percent across the entire USAF.

Figure 3.1. Infrastructure Cost Estimation Approach and Sources

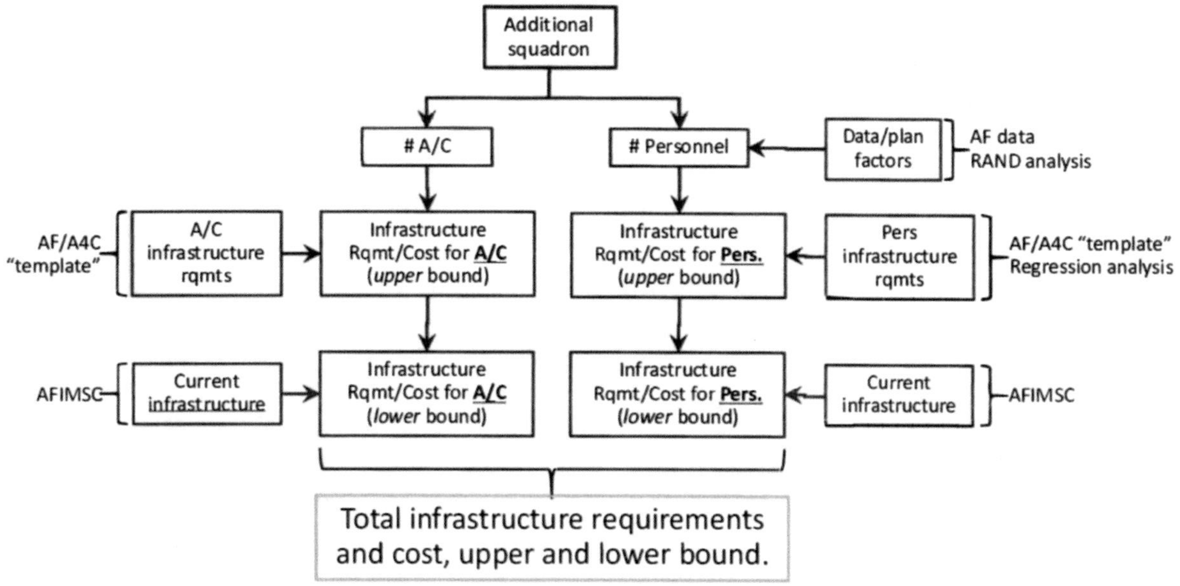

From the total aircraft quantities, we applied AF planning factors that estimate the quantity of infrastructure in various categories required to support each aircraft type. The primary source of these data was the "Civil Engineering Planning Template"[30] (CE template), an unofficial product that primarily draws from AFMAN 32-1084 (2016). This data source encompassed all of the major categories of aircraft-driven infrastructure and facilities (e.g., parking aprons, maintenance facilities, storage).

For the personnel-driven infrastructure categories, we drew, in part, from the CE template, but we predominantly derived regression analyses from personnel data sets. We explain this later in this chapter.

We then applied unified facilities criteria (UFC) cost factors to generate the total cost of each category of infrastructure needed, and then the aggregate cost of new infrastructure per squadron.[31] The aggregate cost, in total, formed an upper-bound cost, which assumes that 100 percent of that infrastructure requirement would require new construction (i.e., zero excess capacity existed at the receiving location to accommodate the incoming forces). We consider this a conservative assumption.

We also sought a lower-bound estimate to be less conservative, and, ideally, more realistic. The AFIMSC already has data and tools to do this type of capacity calculation.[32] With these, it

[30] Provided to authors by AF/A4C on January 16, 2019, via email.

[31] See U.S. Department of Defense (DoD), 2018. Specifically, we drew infrastructure type cost factors (construction, sustainment, and life cycle) from Table 3 and area cost factors from Table 4.1.

[32] AFIMSC, Installation Support Directorate performs this analysis with the help of contractors.

frequently performs analyses for the USAF to take into account (1) the current excess capacity (or deficit) at a given location to accommodate incoming forces and (2) the undeveloped but developable land on the base on which new infrastructure could be constructed to further expand capacity. We provided AFIMSC with a limited set of bases of interest (after much of our effectiveness analysis had been completed) for specific capacity calculations to provide a limited set of illustrative cases for comparison with our baseline upper-bound estimates. We discuss the results of this analysis in the "Capacity-Informed Infrastructure Cost Estimates" section later in this chapter.

In the rest of this chapter, we refer to our initial capacity-ignorant estimate as our *baseline costs* and the second as *capacity-informed.*

Baseline Infrastructure Cost Estimate

The baseline infrastructure cost estimate is based on an analysis of major categories of aircraft-driven infrastructure and facilities (e.g., parking aprons, maintenance facilities, storage). Figure 3.2 illustrates the A4C-provided template that contains these data. The template contains nearly every MDS currently in the USAF inventory (36 total, including several nonflying systems). It then lists more than 70 attributes of those aircraft, including 34 infrastructure categories (called *category codes* [CATCODEs]), such as runway, parking apron size, and jet fuel storage.[33] In Figure 3.2, we have truncated the template significantly horizontally and vertically to make it more legible. The template lists each MDS, the standard squadron PAA size, and the per-squadron requirement for each infrastructure category, in the respective units (e.g., square feet [sq. ft.]). SMEs with whom we spoke acknowledged known shortcomings in this template.[34] The CE template is the best source of aircraft-driven infrastructure requirements, and we used the template as is, resolving only places where it had blank spaces or something that seemed particularly out of order (e.g., using like systems as analogs).[35]

---

[33] A CATCODE is a five- or six-digit code used by DoD that represents a specific type of facility. CATCODEs are assigned in accordance with DoDI 4165.03, 2015.

[34] Conversations with ACC/A4 and ACC/A8 on May 8, 2019, and AFIMSC, Installation Support Directorate on May 21, 2019. What we surmised from these conversations is that, because of the diversity of data in the template, there is no one "owner" of the requirements in that document who would be responsible for updating it. It draws from several sources, both formal and informal. Those with whom we discussed it felt that no contributing organization wanted to be fully responsible for the entire product or even the part that represented the organization's own requirements. To the extent that those individual requirements find their way into formal documentation, formal processes already exist to keep them updated. Despite the document's obvious utility (numerous organizations are aware of it and use it in some form), it remains in this half-living state.

[35] The F-35, in particular, had numerous blank spaces; in most cases, we used the F-22 as an analog.

Figure 3.2. Civil Engineering Infrastructure Template

	ASSUMPTIONS						FACILITY REQUIREMENTS					
Weapon System / Aircraft Type	Minimum PAA	Typical PAA per Sqdn	Load Bearing Strength	Aircraft Tail Height (ft)	Aircraft Wing Span	Aircraft Length	Minimum Runway Length (ft)	Minimum Runway Width (ft)	Minimum Taxiway Width (ft)	Apron (sq yd per aircraft)	Squadron Ops (sq ft)	Flight Simulator (sq ft)
				TAIL			111-111L	111-111W	112-211W	113-321	141-753	171-212
Airborne Laser (ABL)	5	5	Max Taxi Weight	64	211.4	231	10600	150	75	13950	65000	15000
A-10	18	24	51000 lbs	14.8	57.6	53.4	8000	150	75	3700	16000	1200
B-1B	12	12	TT450	34	137	146	12000	150	75	8250	35000	18000
B-2A	8	8	TT336.5	17	173	70	10000	150	75	7900	18000	36000
B-52	12	12	TT-500	41	185	159.4	12000	200	175	10750	35000	18000
E-3 (AWACS)	4	6	ST172	42	145.8	153	7000	135	75	8800	33000	13000
E-4 (NAOC)	3	3	DTT	63.5	196	231	12000	148	75	17600	40000	0
E-8 (JSTARS)	2	5	TT	42	145.8	153	9000	135	75	8800	32000	13500
EC-130H Compass Call	5	5		38.3	132.7	97.9	7000	150	50	7100	33000	5000
F-15C	18	24		18.8	42.1	63.9	8000	150	75	3250	16000	8400
F-15E	18	24		18.8	42.1	63.9	8000	150	75	3250	16000	8400
F-16	18	24		16.4	32.5	49	8000	150	75	2750	16000	8400
F/A-22	18	24	TT83	16.6	44.5	62	8000	150	75	3350	18600	22000
F-35 (JSF)	18	24		15	35	50.5	8000	150	75	TBD		

Applying the UFC cost factors to generate the total cost of each category of infrastructure needed and then aggregating for each squadron type results in aircraft-driven infrastructure costs ranging from $63 million to $92 million, as shown in Table 3.2.

Table 3.2. Baseline Aircraft-Driven Infrastructure Costs

MDS	Aircraft-Driven Infrastructure ($M)
A-10	68
F15C	64
F15E	65
F-16	63
F-22	92
F-35A	92

NOTE: Values are in 2019 dollars.

To identify infrastructure categories driven by base population, we also started with the CE template. However, there were some categories for personnel-driven infrastructure not included, so we extended the template to consider any additional categories not driven by aircraft or by base characteristics. What remained was a list of several dozen infrastructure category candidates for inclusion in our cost model.

We then performed a regression of infrastructure capacity and personnel to determine a relationship between personnel and these infrastructure categories. Eight of the infrastructure categories had a reliable relationship with personnel that we used to develop a mathematical relationship for use in our cost model. Those categories include general administration building, warehouse, dispensary and clinic, child development center, fitness center, dental facility, medical warehouse, and school-age child care. We then applied the UFC cost factors to these categories to generate a cost-per-person figure in each category, then scaled up to a cost per squadron. Finally, we aggregated each individual infrastructure category cost to determine the total per-squadron cost. This results in a personnel-driven infrastructure cost of about $25.3 million per squadron. Table 3.3 shows the resulting cost estimate. The top three categories, general administration building, warehouse, and dispensary and clinic, account for nearly 90 percent of total personnel-driven infrastructure costs.

Table 3.3. Personnel-Driven Infrastructure Costs

Building Category	Cost per Squadron
General administration building	$14,273,200
Warehouse	$4,373,556
Dispensary and clinic	$3,849,904
Child development center	$765,084
Fitness center	$677,332
Dental facility	$604,440
Medical warehouse	$466,036
School-age child care	$252,960
Total	**$25,262,512**

Table 3.4 brings all of these costs together. The MDS-specific aircraft-driven costs are in the first column, ranging from $63 million for fourth-generation fighters to $92 million for fifth-generation fighters. The personnel-driven costs are approximately $25 million per squadron. Although manning numbers do vary across MDS (e.g., two engines versus one, or air-to-ground versus air-to-air-only munitions), we use average squadron size numbers to simplify the calculations. The personnel move costs are $13 million per squadron. Thus, the total one-time (upper-bound) cost associated with squadron moves ranges from $101 million to $130 million per squadron.

Table 3.4. Baseline Squadron Relocation Cost Summary

MDS	Aircraft-Driven Infrastructure ($M)	Personnel-Driven Infrastructure ($M)	Personnel Move ($M)	Total Cost ($M)
A-10	68	25	13	106
F-15C	64	25	13	102
F-15E	65	25	13	103
F-16	63	25	13	101
F-22	92	25	13	130
F-35	92	25	13	130

NOTE: Values are in 2019 dollars.

Capacity-Informed Infrastructure Cost Estimates

At the request of the project team, AFIMSC provided data on Edwards AFB (California), Eielson AFB, Hill AFB, Joint Base Langley-Eustis, and Mountain Home AFB. These bases were selected for additional analysis because they are good candidates for receiving additional units based on their proximity to ranges likely to be upgraded (per AF/A3T's then-current plan).

Table 3.5 shows the surplus or shortfall for each infrastructure category, scaled to squadron equivalents (amount of infrastructure divided by requirement for a single squadron). We see that there is diversity across locations and categories as to whether there is a current surplus or deficit, and how much of each. Most of these bases appear to have a significant surplus in parking spaces and jet fuel storage, but most of the other surpluses and shortfalls are less than one squadron.

Table 3.5. Infrastructure Capacity (Squadron Equivalents) for Five Example Bases

Category	Edwards (F-16)	Eielson (F-16)	Hill (F-35)	Langley (F-22)	Mountain Home (F-15)
Parking spaces	8.1	4.9	6.8	0.0	4.8
Squadron operations	-3.7	0.0	0.3	0.0	1.1
Field maintenance training	-0.1	-0.2	-0.1	0.0	-0.1
Flight simulator	-0.1	-0.3	-0.1	-0.1	-0.1
Small aircraft dock spaces	-0.3	0.3	0.0	0.0	0.1
General purpose shop	0.0	-0.1	4.6	0.0	-0.1
Aircraft maintenance unit	-0.1	-0.1	0.1	-0.1	0.1
Engine maintenance	2.0	-1.6	0.2	0.0	0.4
Munitions production	0.0	0.0	0.0	0.0	0.0
Vehicle maintenance	0.0	0.0	0.0	0.0	0.0
Weapons release	-0.1	-0.1	0.0	-0.2	0.0
Conventional munitions shop	0.0	-0.2	0.0	-0.1	0.0
Avionics shop	0.0	0.0	-0.2	0.0	0.0
Support equipment shop/storage	0.1	0.2	0.7	0.1	0.1
Jet fuel storage	225.4	1,359.2	0.0	38.3	82.3

NOTE: A negative number indicates a shortfall.

In our analysis, we treat the surpluses and shortfalls differently. For each base, we considered the shortfall as a total "debt" to be paid before estimating new construction costs. This means that, before a new squadron could be accommodated, the existing deficits must be covered so that at least the existing forces can be adequately supported. Figure 3.3 shows the infrastructure shortfall, in terms of dollars, for each base, broken out by infrastructure category. Most of the deficits fall within three CATCODEs: field maintenance training, flight simulator, and weapons release. Most of the bases have significant shortfalls in all three categories. The total deficits range from $28 million (Hill) to $115 million (Eielson).

Figure 3.3. Infrastructure Shortfall in U.S. Dollars

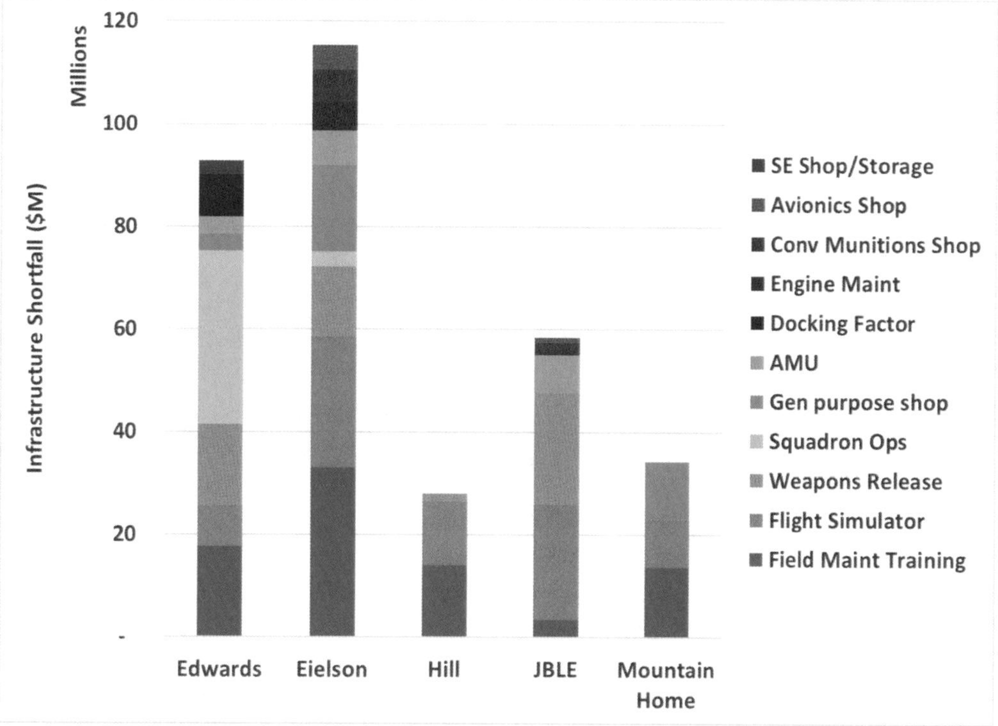

NOTE: AMU = aircraft maintenance unit; Conv = conventional; Gen = general; JBLE = Joint Base Langley-Eustis; Maint = maintenance; Ops = operations; SE = support equipment.

When we incorporate the surpluses, we simply subtract the number of additional squadrons from the squadron-equivalent surplus in each infrastructure category. If the surplus exceeds the new demand, there is no additional infrastructure cost; if the surplus does not cover the new demand, we estimate the investment cost of the remainder. We then aggregate all of the categories with additional infrastructure investment costs.

Figure 3.4 shows the net result of all of the prior calculations. Here, the gray columns show the range of capacity-informed costs across the five bases analyzed. The leftmost column ranges from $28 million to $115 million, the aggregated base-level deficits discussed previously. Averaging across the five bases results in about $76 million in fixed costs to add a squadron to a base. Each gray column to the right is the variable cost of adding an additional squadron and incorporates the surplus calculation, such that the per-squadron cost is less than our baseline cost. The average variable cost per squadron, incorporating surpluses, across all bases, is about $50 million per squadron, roughly half of our baseline costs. The blue and black lines represent our baseline infrastructure costs per squadron. The baseline estimate has a fixed cost (y-intercept) of zero and a variable cost (slope) of either $100 million or $130 million, for fourth- or fifth-generation fighters, respectively, while the capacity-informed costs are fixed at about $76 million and the variable cost at about $50 million.

Figure 3.4. Capacity Summary

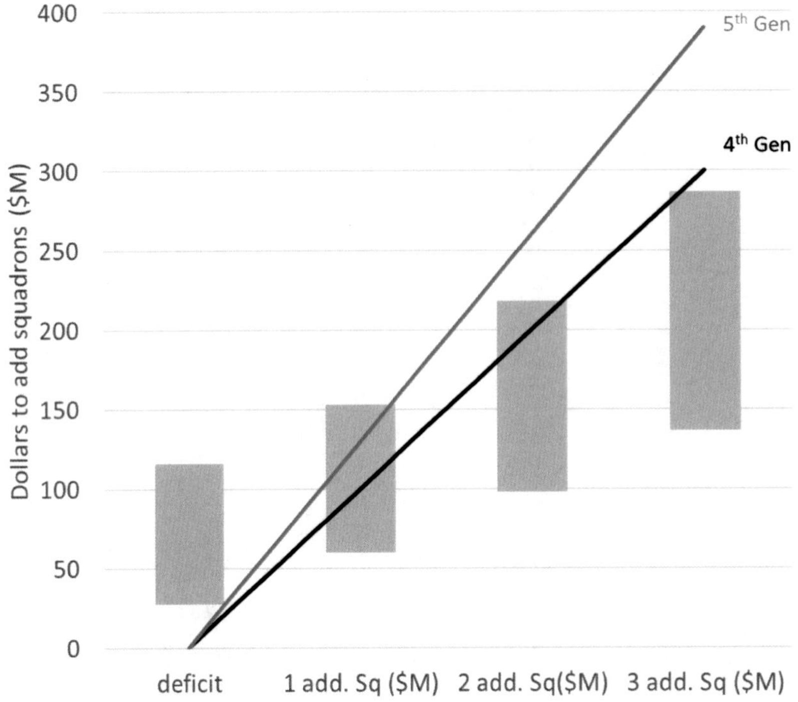

NOTE: Add. = additional; Gen = generation; Sq = squadron. Black line shows baseline (capacity-ignorant) per-squadron cost of fourth-generation fighters; blue line shows same for fifth-generation fighters. Gray columns represent entire range of six example bases; lower extreme is minimum value; upper extreme is maximum value.

We see that for the one added squadron, the range of capacity-informed estimates entirely overlaps the fourth- and fifth-generation baseline estimates. In other words, our original baseline estimates are at least in line with these capacity-informed numbers for this small sample size. In the case of adding two squadrons, our baseline estimates are a bit high, either at the upper end of the range or a bit above it. Thus, if more than one new squadron is added, the example bases we used would have sufficient surplus capacity (given the resolution of the data and tools used by AFIMSC) to significantly discount our baseline estimates, such that our baseline costs applied across the board would noticeably overestimate the one-time investment costs. Because we constrained our analysis to adding only one or two new squadrons, our cost estimates seem reasonable, given the resolution of this part of our analysis. However, future analyses could result in greater fidelity on specific cases to make more direct cost-effectiveness trade-offs.[36]

---

[36] One element that we did not explore is the amount of developable land available at each location. That is an important element in estimating the actual capacity to build new infrastructure. AFIMSC does have data on developable land and uses it in its own analysis for USAF customers.

Range Upgrade Costs

As discussed in Chapter 2, AF/A3T is developing an OTI investment plan to upgrade certain ranges with sufficient capabilities to provide fighter pilots with advanced training. Table 2.3 presented the prioritized list of range upgrades in the current AF/A3T plan. In addition to developing a range upgrade priority, AF/A3T has developed preliminary cost estimates for the research and development and procurement for the capabilities required to upgrade each range.

AF/A3T estimates that $1.2 billion will be required for research, development, test, and evaluation (RDT&E) funding. This is the cost required to develop the representative threat systems and is independent of the number of systems ultimately fielded. As discussed previously, two ranges, the NTTR and JPARC, will be upgraded with the most-advanced capabilities to support large force exercises and were not within the tradespace analyzed in this study. AF/A3T estimates that procuring equipment to upgrade these two ranges will cost roughly $1 billion. The remaining 15 ranges in the upgrade plan are set to be upgraded as primary training ranges with the cost for procurement varying from about $120 million to about $220 million (averaging about $165 million) per range depending on existing and desired capabilities at each range. This includes just procurement costs, not recurring costs. There are recurring costs to operate and sustain range equipment and infrastructure, as well as future modernization costs to continue to increase capabilities as adversary capabilities grow. These could be significant, but were not available at the time of this analysis. The one-time costs are summarized in Figure 3.5. All costs were provided by AF/A3T and were not independently validated by us.

Figure 3.5. Range Upgrade Cost Estimates

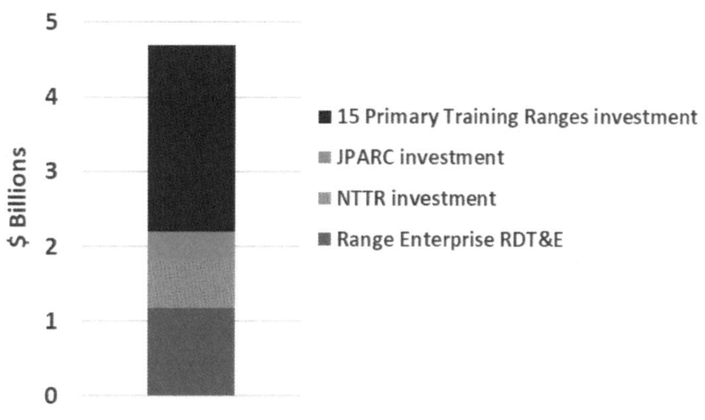

Comparing Restationing and Range Upgrade Costs

As the strategy for providing access to advanced training develops, the USAF may want to consider the cost-effectiveness of different solutions. A robust cost-effectiveness analysis requires full life cycle cost estimates, which were beyond the scope of this study. However, comparing the one-time cost associated with squadron restationing and range upgrades can provide some initial insights and lay the foundation for such an analysis. Using the analysis presented in this chapter, the restationing cost for one squadron ($101 million to $130 million) is on the same order of magnitude as a primary training range upgrade (about $165 million). However, upgrading a single range may provide access for more than one squadron, and a cost-effectiveness assessment should be conducted that accounts for the life cycle range modernization costs.

4. Risk Analysis

The primary focus of this project was an analysis of how various range upgrade and squadron restationing policies interact to provide certain levels of *effectiveness* as described in Chapter 2. Subsequently, in Chapter 3, we compared the one-time costs associated with range upgrades and squadron restationing to lay the foundation for assessing potential *cost-effectiveness* trade-offs between those options. In this chapter, we introduce a third consideration, *risk* from natural hazards, climate effects, and power disruptions that could affect anticipated effectiveness or cost associated with a particular basing decision. Moving forces, equipment, or operations to a base or range with low power reliability or increased susceptibility to operationally relevant climate effects can mean that the actual effectiveness is lower than the anticipated effectiveness. Similarly, the life cycle costs, including those associated with risk mitigation or recovery, ought to be considered as part of the decision. Accounting for such risks, even if it is not possible to precisely quantify them, can provide a more comprehensive view of the benefits and costs associated with basing decisions. Although we did not include risk in our effectiveness or cost models in the previous chapters, the concepts discussed in this chapter begin to lay a foundation for incorporating risk.

A brief survey of recent events at AFBs highlights the importance of including risks in basing decisions. Vandenberg AFB in California experienced disruptions to operations because of a wildfire in September 2016. Although there was no physical damage to the base, the wildfire spread within the vicinity of two space launch pads on site and resulted in the delay of a scheduled rocket launch (DoD, 2019). Hurricane Michael caused significant damages in fall 2018 at Tyndall AFB in Florida. In March 2019, several buildings at Offutt AFB in Nebraska were affected by flooding of the Missouri River. The damages to these bases and the cost of rebuilding them are estimated at approximately $4.7 billion and $700 million for Tyndall and Offutt, respectively (Gould, 2019). Other climate-related effects, such as increases in extreme temperatures, can affect the performance of aircraft (e.g., by imposing restrictions on takeoff weight, runway length) or people (e.g., by reducing the amount of time that can safely be spent outdoors). These examples show that natural disasters and other disruptions can impede military operations and have significant financial consequences for the USAF. Although it is difficult to predict when or how often these events will occur, the impact can be significant. Planning for them now using available information could reduce USAF mission impact, spending on post-disaster rebuilding, or both.

Toward that end, we gathered publicly available data and information, including hazard exposure maps, downscaled climate projections, and historical electric grid reliability data, that can provide high-level insights into the relative susceptibility of different USAF bases and ranges to different types of hazards and threats. Ultimately, when it comes to understanding specific mission- or installation-level vulnerabilities and capacities for coping with disasters, there is no substitute for localized, in-depth assessments that account for a range of disruption scenarios. Also, risk cannot be eliminated. It can only be made explicit, so that decisionmakers can choose how much risk to "buy down" or accept. However, in pooling together these publicly available data and model outputs, we aim to take a step toward systematically incorporating these considerations in basing decisions.

Hazard Exposure Maps

Hazard exposure maps provide a useful view of the varying susceptibility of different geographic regions to different types of natural disasters and other climate-related events. We collected National Oceanic and Atmospheric Administration (NOAA) sea level rise (SLR) inundation maps, which illustrate the scale of potential coastal flooding, for different SLR scenarios (e.g., 3 feet, 7 feet, 10 feet) (NOAA, Office for Coastal Management, undated-a; NOAA, Office for Coastal Management, undated-b; NOAA, Office for Coastal Management, undated-c). We also examined U.S. Forest Service (USFS) Wildfire Hazard Potential maps that present a qualitative interpretation of wildfire risk using historical data and land use. Finally, we reviewed the U.S. Geological Survey's (USGS) seismic hazard maps that show the relative potential for earthquakes in a particular area.

Overlaying fighter bases and ranges on top of the hazard exposure maps can identify the different risks associated with each. For example, Figure 4.1 shows the SLR-induced inundation that would result at Joint Base Langley-Eustis and Tyndall AFB. Exposure information for all bases and ranges considered in this study is included in Table B.3 in Appendix B. Figure 4.1 provides an example view of the types of hazard exposure data that could be visualized for selected locations.

Figure 4.1. Projected SLR Inundation for Joint Base Langley-Eustis and Tyndall AFB for 3-Foot SLR Scenario

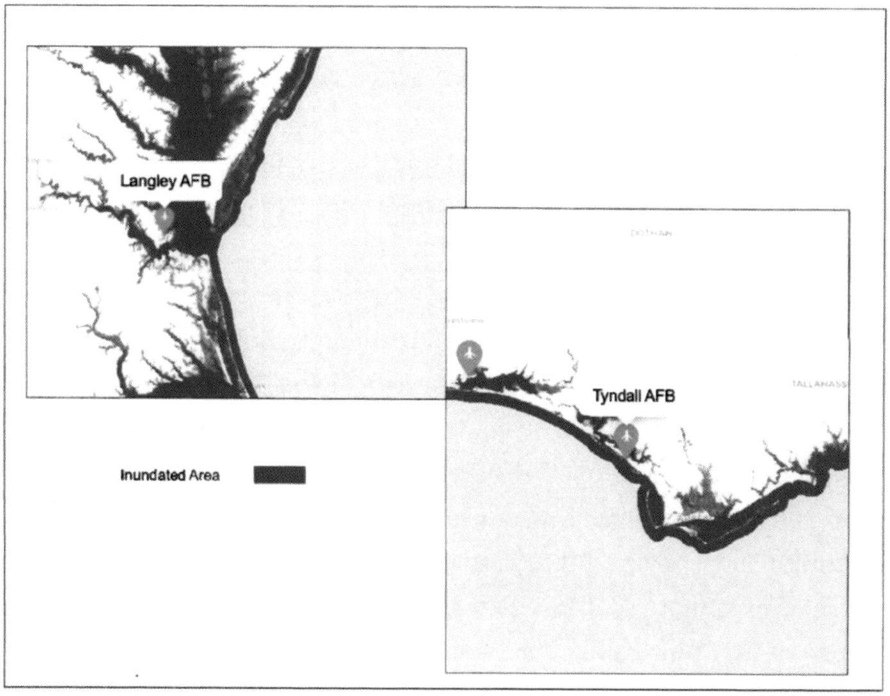

SOURCE: Authors' analysis of NOAA SLR inundation maps from NOAA map server files (NOAA, Office for Coastal Management, undated-a).

There are important caveats to consider when using hazard exposure maps to understand risk. First, there are differences in the approach taken to generating different types of hazard exposure maps. These differences have implications for how the information should be interpreted and used for planning purposes. Hazard maps for SLR are not associated with a probability distribution or temporal scale. Instead, they represent a range of potential future outcomes, each of which is calculated using a different set of parameterized assumptions about the future. Therefore, there is no time frame inherently associated with the different SLR scenarios used to show inundation at a given location. The wildfire hazard exposure map illustrates the potential for wildfire based on five wildfire hazard potential classes (very low to very high) and two nonwildfire hazard potential classes (nonburnable and water). This exposure is also not linked to a particular time frame. Finally, the seismic hazard maps, on the other hand, show horizontal spectral response acceleration with an associated exceedance probability (10 percent) over a specified period (50 years). There may also be geospatial variability in some of these future outcomes. SLR, for example, may not affect all points on a coastline under a similar time horizon. Hazard exposure maps that include projections on the future do not incorporate different scenarios of a changing climate over time.

Downscaled Climate Projections

Climate projections are simulations that attempt to provide information on future climate conditions using such variables as air temperature, precipitation, relative humidity, and wind speed. These simulations, which are called General Circulation Models (GCMs), use complex mathematical equations that simulate Earth's land, oceanic, and atmospheric systems.[37] Unlike weather forecasting, climate projections span larger geographical areas and longer periods. Each GCM is typically modeled at a grid size of hundreds of kilometers and over a time scale of decades (NOAAClimate.gov, undated).

To compare future climate conditions for different USAF bases, we used downscaled climate projections.[38] For every CONUS fighter base and range, we downloaded downscaled climate projections on a daily time step from the Northwest Knowledge Network's Multivariate Adaptive Constructed Analogs (MACA) data set (Northwest Knowledge Center, undated). Specifically, we gathered outputs of 12 different GCMs for the variables listed in Table 4.1 for the period from 2006 to 2099.[39] Each of the 12 GCMs uses different assumptions and mathematical formulations that result in variability across model outputs. There is no evidence to suggest that one GCM is more accurate than another, and using climate projections from various GCMs provides a variety of possible outcomes for the climate variables under consideration.

---

[37] Climate projections vary based on assumptions regarding greenhouse gas emissions. The Intergovernmental Panel on Climate Change established a set of scenarios called *Representative Concentration Pathways* (RCPs) to account for different amounts of radiative forcings on Earth as a result of greenhouse gas emissions in the atmosphere. The RCPs (measured in watts per square meter) are categorized in four groups: RCP 2.6, RCP 4.5, RCP 6, and RCP 8.5. The RCP 4.5 scenario represents a more positive outlook on reducing climate change impacts in the future where greenhouse gas emissions are reduced from their current levels. The RCP 8.5 scenario represents a business as usual case where greenhouse gas emissions stay consistent with current trends.

[38] Projections are typically downscaled using one of two methods: dynamic downscaling or statistical downscaling. Dynamic downscaling involves running simulations at a higher resolution by incorporating such local conditions as topography and land features at a higher fidelity into the model. Statistical downscaling involves using global climate models at a lower resolution in combination with statistical relationships between local and global climate systems to output higher-resolution projections (Hannah, 2015). Statistical downscaling certainly comes with pitfalls (see, for example, Lanzante et al., 2018). The approach relies on comparing and reconciling differences between empirical observations and outputs of physical models for a time period in the past (training step), and then applying the "corrected" model to a future time period. An underlying assumption is that fundamental attributes of climate will not change too much in the future. This assumption is difficult to validate. Nonetheless, statistical downscaling is commonly used in the climate science community and can still provide useful insights if model outputs are interpreted carefully.

[39] See Appendix B, Table B.1 for GCM references.

Table 4.1. Description of Variables in Climate Projections

Climate Projection Variable	Description
Near-surface specific humidity	Daily average ratio of water vapor mass per unit mass of air (kg/kg)
Precipitation	Daily rainfall (mm)
Minimum near-surface relative humidity	Daily minimum relative humidity (%)
Maximum near-surface relative humidity	Daily maximum relative humidity (%)
Surface downwelling solar radiation	Daily average surface downswelling radiation (w/m2)
Minimum near-surface air temperature	Daily minimum near-surface temperature (°C)
Maximum near-surface air temperature	Daily maximum near-surface temperature (°C)
Eastward component of wind at 10m	Daily average eastward component of wind, measured at 10m (m/s)
Northward component of wind at 10m	Daily average northward component of wind, measured at 10m (m/s)

We calculated additional metrics using the outputs listed in Table 4.1. These additional metrics, as shown in Table 4.2, were calculated to provide further insight into climate change projections using the raw downscaled data.

Table 4.2. Description of Metrics Generated Using Climate Projections

Climate Projection Variable	Description
Hundred-degree days	Mean annual number of days over a given time period where the daily maximum air temperature exceeds 100°F
Mean annual number of black flag days	Mean annual number of days over a given time period where the estimated wet bulb globe temperature exceeds 90°F. Estimated using maximum daily temperature (*tasmax*), solar radiation (*rsds*), and mean daily near-surface relative humidity (*rhsmin* and *rhsmax*)
Mean specific humidity	Mean specific humidity, calculated using daily MACA data (kg/kg)
Mean specific humidity, dry season	Mean specific humidity (kg/kg) during the driest average three-month period over specified time frame (location specific)
Mean specific humidity, wet season	Mean specific humidity (kg/kg) during the wettest average three-month period over specified time frame (location specific)
Mean precipitation, dry season	Mean precipitation (mm) during the driest average three-month period over specified time frame (location specific)
Mean precipitation, wet season	Mean precipitation (mm) during the wettest average three-month period over specified time frame (location specific)
Maximum KBDI during hottest three months	Mean annual maximum estimated KBDI during average hottest three-month span (location-specific)
Mean annual precipitation	Mean annual total precipitation over specified time frame (mm)
Average daily mean temperature	Average daily mean temperature over specified time frame (°C)
99th percentile (all metrics)	99th percentile of a selected metric over the specified future time period

NOTE: KBDI = Keetch-Byram Drought Index.

As an example, Figure 4.2 depicts the annual average maximum temperature for Hill AFB for 12 GCMs under RCP 8.5 for the years 2025 to 2085. Although there is variability in the projections for the 12 GCMs, they all indicate a general upward trend with maximum annual average temperatures for Hill AFB reaching 66 to 78 degrees Fahrenheit toward the end of the century.

Figure 4.2. Projected Annual Average Maximum Temperature for Hill AFB Under RCP 8.5

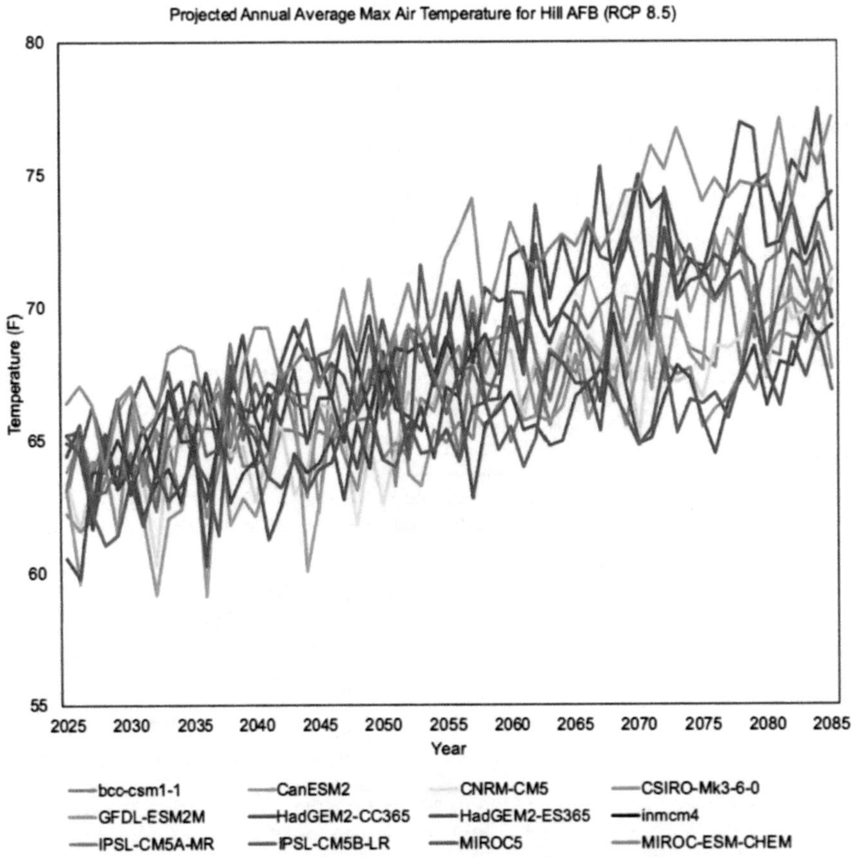

NOTE: See Appendix B, Table B.1 for GCM references.

A perpetual challenge in making use of climate data is the ability to link it to operational impacts. Climate-related events, such as hurricanes, wildfires, and flash flooding, have led to substantial USAF operational impacts, but it is difficult to attribute climate change projection data to specific climate-related and natural hazard events. For example, projections in precipitation levels alone are not enough to predict whether a USAF base or range will experience flash flooding from a storm surge. Instead, additional modeling at a regional level and information on intensity of precipitation, drainage conditions on site, and infrastructure design characteristics would be needed to make informed projections of flash flood risk. Thus, climate projection data may be most useful as inputs for conducting in-depth and localized vulnerability assessments to understand the operational impacts of climate-related effects for the USAF.[40]

Conducting such assessments was outside the scope of this project, but we did investigate one potential link between climate data and operational considerations for basing decisions. The USAF has a thermal injury prevention program that outlines precautions required when performing duties in extreme temperatures (AFI 48-151, 2016). That program includes the definition of *black flag conditions*, indicating “extreme” heat stress risk, occurring when the wet bulb globe temperature (WBGT), a measure of heat stress, is higher than 90 degrees Fahrenheit.[41] In such conditions, moderate (e.g., light maintenance work) to hard (e.g., loading and unloading pallets) work is limited to 10–20 minutes per hour compared with 40–60 minutes per hour in cooler weather conditions. As a result, an increase in the number of black flag days could reduce operational capability. Figure 4.3 shows the expected average increase in black flag days at Tyndall AFB in the near term for RCP 4.5 and RCP 8.5.[42] Each dot represents a model output (with results from 12 GCMs shown in the figure). The horizontal gray bar shows the median value for each RCP.[43] Results for all bases and ranges are in Appendix B, Table B.4.

---

[40] Although it may be possible to account for some operational impacts, it is not likely that these data will be able to capture all mission-related weather phenomena.

[41] The WBGT is calculated using the natural wet bulb temperature, the black globe temperature, and the dry bulb temperature. The dry bulb temperature input in the WBGT equation is one of the direct outputs (maximum air temperature) of the MACA climate projections. We estimated the natural wet bulb temperature and the black globe temperature using existing empirical models that have been developed by the scientific research community. The model used to estimate the wet bulb temperature incorporates climate projection data on relative humidity and maximum air temperature (Stull, 2011). Note that the regression is based on data gathered at sea level, so the calculation may be less reliable at higher altitudes. The model used to estimate the black globe temperature incorporates climate projection data on maximum air temperature and solar irradiance (Hajizadeh et al., 2017).

[42] The *near-term period* is defined as the years 2026 to 2055.

[43] Note that confidence in downscaled projections for relative humidity is limited, so WBGT projections should be used only to gain a general understanding of future trends.

Figure 4.3. Projected Increase in Mean Annual Number of Days with Black Flag Conditions at Tyndall AFB for Near Term

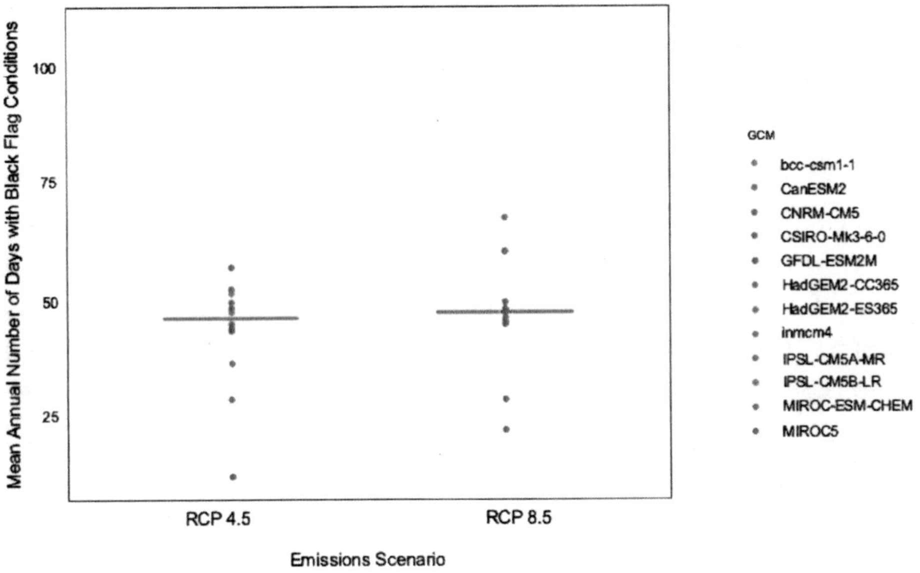

Electric Power Reliability

The extent to which USAF installations and training ranges maintain uninterrupted power service depends on both the *reliability* of the electric power supplied from the utility company and the *resilience* measures implemented inside the fence line of base and range installations.[44] On-site power resilience measures (such as diesel generators) may make up for some lags in grid reliability (Narayanan et al., 2017). An important distinction between the reliability of electric power supply from the utility company and implementation of resilience measures inside the fence line is that the USAF tends to directly control only the latter. Thus, in areas where electric power supplied by utility companies is less reliable, the USAF may experience more outages or require additional investment in power resilience measures.

Electrical grid–related power disruptions vary across the United States and can be considerable. In 2017, electric customers in the United States experienced an average duration of 7.8 hours of electricity interruptions including major event disruptions and approximately 1.9 hours of electricity interruptions excluding major events (U.S. Energy Information Administration, 2018). We use three performance metrics to characterize the reliability of grid power: System Average Interruption Frequency Index (SAIFI), System Average Interruption Duration Index (SAIDI), and Forced Outage Rate (FOR). Our analysis does not account for inside-the-fence resilience measures.

---

[44] *Reliability* is "the confidence in the actual power characteristics provided to a point in the system," and *resilience* is "the ability of a system to withstand and recover from a disruption," as defined by Narayanan et al., 2017.

The SAIFI is a measure of how often an electrical utility customer, on average, experiences a major power disruption (a disruption longer than five minutes) and is measured as the number of such power disruptions (i.e., the total number of customers who experienced a disruption divided by the total number of customers). The SAIDI is a measure of how long (in minutes) an electrical utility customer, on average, experiences a major power disruption (i.e., the total number of minutes of disruption divided by the total number of customers). Finally, the FOR is the percentage of time that a power generating unit will not be available for service (i.e., the total number of hours a unit is on a forced outage divided by the total number of hours in a year) (Resource Adequacy Planning, 2017). Taken together, these metrics provide some insight into the reliability of the power grid in terms of number of disruptions, duration of disruptions, and likelihood of unscheduled disruptions.

SAIFI and SAIDI are performance metrics that utilities report to the U.S. Energy Information Administration (EIA) through EIA Form 861.[45] We collected SAIFI and SAIDI data for the years 2013 to 2017 and averaged across years for each respective utility and then used utility service territories as a rough way to map a given base or range to a particular utility.[46] Historical FOR values for power generating units are made available through the North American Electric Reliability Corporation (NERC) Generating Availability Data System for power generating units of varying capacity (in megawatts) and generation technology (e.g., coal, natural gas, nuclear) (North American Electric Reliability Corporation, undated). We provide a note of caution in using SAIDI or SAIFI to arrive at an estimate of the reliability of commercially provided electric power at individual bases: For bases that fall in larger service territories, the reliability of commercial power may appear higher or lower than it is in reality. This is because the smallest unit of reporting is the utility service territory, and reliability values for the territory represent an average over the entire customer base served by the utility. That being said, although individual bases might have a better idea of the localized reliability of the commercial grid, we were unable to access any centralized repository of base-specific power reliability data. In our quest to provide an enterprisewide view of grid reliability, we were limited to broadly available metrics, such as SAIDI and SAIFI.

Unlike SAIFI and SAIDI, the FOR values for each utility are not reported directly, so we calculated an average FOR value for each utility based on the generator size and generation technology of power plants serving that utility. By mapping the three performance metrics across bases and ranges, we can see the heterogeneity in electric grid reliability as characterized through average SAIDI, SAIFI, and FOR. As an example, Figure 4.4 depicts the average SAIDI values

---

[45] EIA, 2017. Utilities in the United States that own electricity generating power plants with a nameplate capacity of 1 megawatt or above report SAIFI and SAIDI annually. The data from the submission of those forms are made publicly available through the EIA website.

[46] SAIFI and SAIDI data for a given utility with available data were averaged for all years from 2013 to 2017. In some instances, data were available for 2013 through 2017, so the average would be over five years. In other instances, data were only available for 2016 to 2017, so the average would be over two years.

for CONUS fighter bases and ranges. The average SAIDI, which is a measure of the average duration in minutes that a customer experiences during an outage, varies from approximately 50 minutes to 1,800 minutes for the USAF installations shown. SAIDI, SAIFI, and FOR for all bases and ranges are in Appendix B, Table B.2.

Figure 4.4. Average SAIDI Value of USAF Bases and Ranges

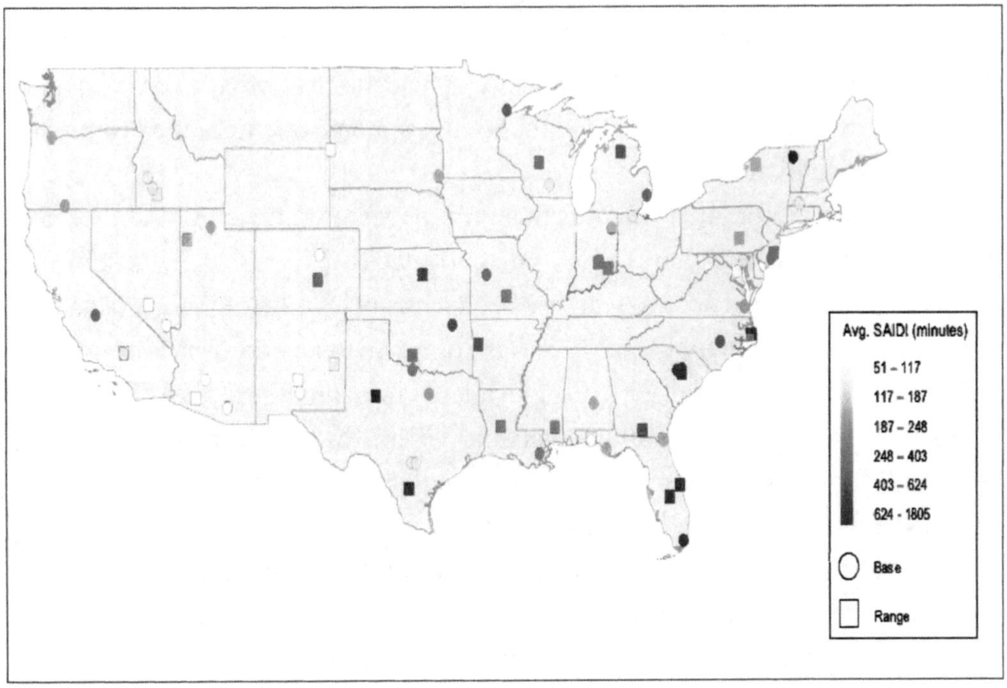

Discussion and Next Steps

The data and methods described in the preceding sections highlight opportunities to better understand geographically based differences and similarities between USAF air bases and ranges, today and in an uncertain future. Drawing on the types of information presented in this chapter, researchers could generate a table, such as Table 4.3, for each base or range that is considered as a candidate destination in a basing action. Additional attributes could be added so that a composite picture of potentially decision-relevant information begins to form.

Table 4.3. Comparison of Risk Considerations Across Tyndall AFB, Hill AFB, and Joint Base Langley-Eustis

Relevant Basing Decision	Bases Considered	Would 3-ft SLR Inundate Location?	Projected Increase in Mean Annual Black Flag Days	Total Number of Hurricane Strikes Within 10 nm (1990–2018)	Historical Average Duration of Power Outage (SAIDI)
Influx of F-35 squadrons	Tyndall AFB Hill AFB	Yes No	~47 days ~4 days	4 0	~230 mins ~250 mins
Influx of F-22 FTU squadron + portion of ops unit from Tyndall	Joint Base Langley-Eustis	Yes	~29 days	1	~200 mins

[a] The value shown is an average across all 12 GCMs for the *near term*, which is defined as a time period between 2026 to 2055 and RCP 8.5.

Even without any additional information on the likelihood of a 3-ft SLR scenario, if the relocation of a USAFR F-35 squadron in isolation is considered, and if both Hill AFB and Tyndall AFB are positioned equally well to meet effectiveness requirements with comparable infrastructure and personnel move costs, then Hill AFB may be the preferred alternative. An alternative interpretation stems from recognizing that squadrons that are displaced as a result of a hurricane may not have anticipated range access, so the *actual* benefits associated with moving to a location (such as Tyndall AFB) that is prone to hurricanes may be lower than the *anticipated* benefits associated with the move.

If other factors are such that Tyndall AFB is still chosen, then, at a minimum, the decision ought to take into consideration the costs associated with mitigating the effects of SLR-induced flooding or a high-intensity hurricane in addition to the restationing costs involved in moving a squadron to Tyndall AFB. Similarly, consolidating F-22 presence at Joint Base Langley-Eustis may incur additional flood mitigation or recovery costs. These costs should be weighed as part of basing decisions. Although it is not possible to predict when the next Category 5 hurricane will hit an AFB, it may be beneficial to consider the question, "How frequently would hazard X need to occur at base Y before the mitigation costs or reduced effectiveness outweigh the factors that led to moving forces and equipment to base Y?"

Finally, there is likely no part of the United States that is hazard-free. However, not every hazard is the same in terms of potential occurrence or impact. The cost of recovering from a Category 5 hurricane that results in significant damages across a large area is not equal to that of recovering from a flood event that affects portions of a base. Understanding the geographically heterogeneous nature of hazard exposure and climate-related effects, through the metrics we have introduced in this chapter, can provide a way to meaningfully differentiate among locations.

There are several steps that the USAF could take to further the analysis described in this report. First, collecting data at a higher resolution can help—highly localized data on topography, design standards, and more are needed to understand how exposure to different external factors is likely to translate to operational impacts. Second, there are several ways to synthesize and visualize the sorts of information described in this chapter. We have begun to develop a data visualization tool that could be improved with input from USAF decisionmakers. Finally, better characterizing operationally relevant metrics as a function of some of the risk-related variables discussed in this report can help decisionmakers contextualize, interpret, and act on outputs of climate projections and grid reliability data.

5. Conclusions

The USAF uses a transparent, repeatable, and defendable process for making individual basing decisions. The analysis presented in this report employed a strategic, portfolio-level assessment of the U.S.-based force posture that shows that the service could achieve significant readiness improvements through restationing of its fighter forces. As a result of this analysis, we make the following conclusions.

Range upgrades alone provide only a portion of all fighter squadrons with access to advanced training ranges. Assuming that all fighter squadrons having access to an upgraded training range represents maximum effectiveness, upgrading all 17 ranges currently identified in the AF/A3T range modernization plan achieves only approximately 70-percent effectiveness. Discussion with SMEs at AF/A3T suggests that only half of those ranges may receive upgrades, which achieves 50-percent effectiveness.

Restationing can significantly increase effectiveness, but the amount depends on which institutional policies the USAF is willing to change or manage. If all ranges are upgraded, ten to 20 squadron moves can increase effectiveness from 70 percent to 90–100 percent. If half of the ranges are upgraded, ten to 20 squadron moves can increase effectiveness from 50 percent to 65–80 percent. However, the potential benefit of restationing depends significantly on the restationing polices the USAF is willing to consider. Our analysis suggests that the largest drivers of increased effectiveness resulting from restationing are ANG squadron consolidation and increased fighter base capacity.

It is too early to advocate for specific basing actions because of training and basing details that still need to be resolved. This analysis showed how combinations of range upgrade and squadron restationing policies achieved various levels of effectiveness. However, it is too early to recommend specific basing actions because of uncertainties related to both training and basing. The USAF is still defining range requirements and capacity to determine how much time would be required at upgraded ranges for each MDS. The USAF will need to consider air-to-air training airspace available in addition to the access to ground ranges. Additionally, full life cycle cost estimates are required to better understand the long-term cost implications of range upgrades. On the restationing side of the equation, a better understanding of base capacity and airspace congestion, both existing and potential, is required to understand the upper limit of fighter consolidation near upgraded ranges and associated costs.

In the near term, our analysis suggests a large potential benefit from consolidating F-22 squadrons near an upgraded range. Our assessment of training requirements suggests that the F-22 is the second-most important MDS to have access to advanced training ranges. However, the F-22s are distributed across bases that are not near ranges targeted for modernization. Moving up the range near Joint Base Langley-Eustis in the upgrade priority ranking and consolidating F-22 squadrons at Joint Base Langley-Eustis (or converting a fourth-generation base near a high-priority range to an F-22 base) would provide a significant increase in effectiveness. In either case, air-to-air training airspace capacity would need to be factored into the decision.

In the longer term, our analysis highlights a large potential benefit of integrating the range modernization plan and the F-35 rollout. Our assessment of training requirements suggests that the F-35A is the most important MDS to have access to advanced training ranges. In addition, because most of these basing decisions have not been made, they may be subject to fewer institutional constraints compared with existing forces. In the coming years, F-35s will enter the operational force, legacy aircraft will be retired, and ranges will be modernized. These decisions can be orchestrated in a manner to ensure that the new force has access to upgraded ranges at the earliest possible time.

The one-time cost for restationing a fighter squadron and the cost to procure equipment for a single primary training range upgrade are on the same order of magnitude. However, when research and development, operational, and sustainment costs are taken into account, range upgrades may be substantially more expensive over the long term. Developing full life cycle cost estimates for range modernization will indicate the number of ranges that are affordable over the long term and enable better comparisons with the cost and institutional challenges of restationing.

There is sufficient variability of electric power reliability and exposure to natural hazards and climate effects across USAF fighter bases and ranges that these factors should be incorporated into basing decisions. Analysis of hazard exposure maps, downscaled climate projections, and electric power reliability showed that the risk to bases and training ranges varies. These publicly available data can be used to make more-comprehensive assessments of the costs and benefits associated with basing alternatives.

Range access will continue to be an issue for the USAF. Restationing, range upgrades, and the associated policies and details are just some of the options for increasing access to advanced training and should be evaluated against other potential alternatives. As discussed previously, even after range modernization, many squadrons may not have local access to advanced training ranges. Squadron restationing is one potential solution, and it was the focus of this analysis. However, using discussions with SMEs and our review of the F-35 and F-22 training requirements, it is reasonable to expect the USAF will need to develop a graduated training strategy in which lower-level training ranges are used to satisfy basic training requirements while highly capable ranges are used for advanced requirements. Depending on the details of such a training strategy, the importance of proximity to ranges—the primary effectiveness metric in this study—could change. For example, if required training at advanced ranges could be accomplished in a few weeks per year, the USAF could consider temporary deployments or the use of tankers to provide range access. Similarly, advancements in integrated LVC capability may allow more simulated training to be done, therefore reducing the requirement for range access. The value of restationing, as assessed in this report, could be evaluated against these other options.

Advanced LVC capabilities may significantly affect the importance of proximity to ranges, but there is still significant uncertainty regarding the development of such capabilities. A network of simulators and instrumented aircraft will be an important component of a graduated training strategy and may significantly reduce the amount of live flying on advanced ranges. The USAF has invested in, and plans continued investment in, the infrastructure necessary to increase the use of LVC in training. Whether these investments ultimately reduce the requirement for close-proximity live ranges depends on a variety of yet-to-be-answered questions, such as the following:

- What is the right mix of live, virtual, and simulator training to prepare CMR pilots for engagements with near-peer competitors?
- Does the USAF have the means to assess the effectiveness of combinations of training method types?
- For LVC training, have key technical challenges been resolved?
- Is the network that connects aircraft and other weapon systems capable of dealing with important security issues?

Without answers to such questions, we are unable to adequately capture the trade-off between LVC and live test and training requirements.

The framework and analysis in this report can inform the Base Realignment and Closure (BRAC) process. BRAC is used by DoD "to reshape the Department's physical plant, that is, its installations and associated weapons ranges, as well as the organization and stationing of its forces."[47] Through BRAC, DoD undertakes rigorous analysis to divest underused infrastructure to reduce annual fixed costs. One of the stated objectives of BRAC is "maximizing both warfighting capability and efficiency."[48] The methods developed in this analysis could be used to help identify bases that are candidates for closure and flying units that are candidates for realignment.

The framework and analysis in this report can inform Air Force Warfighting Integration Capability (AFWIC) analysis of future force design. AFWIC is responsible for conducting integrated analysis to support future force design. An extension of the framework described in this report could be used to assess the impact of different future combat forces on the availability of ranges to support training requirements. If force design alternatives drive the need for additional upgraded range capacity, the associated costs should be captured and included as part of the force design trade-off analysis that AFWIC uses to inform senior decisionmakers.

---

[47] DoD, 2005, p. 1.

[48] DoD, 2005, p. 3.

Appendix A. 2025 Fighter Force Structure

Table A.1. 2025 Force Structure

Squadron	Component	Command	Current Base	MDS
119th FS	ANG	ANG	Atlantic City International Airport, Egg Harbor Township	F-16
131st FS	ANG	ANG	Barnes ANGB, MA	F-15C
120th FS	ANG	ANG	Buckley AFB, CO	F-16
134th FS	ANG	ANG	Burlington ANGB, VA	F-35A
357th FS	RegAF	ACC	Davis-Monthan AFB	A-10C
47th FS	AFR	AFRC	Davis-Monthan AFB	A-10C
354th FS	RegAF	ACC	Davis-Monthan AFB	A-10C
121st FS	ANG	ANG	DC National Guard, JB Andrews, MD	F-16
179th FS	ANG	ANG	Duluth ANGB, MN	F-16
416th FLTS	RegAF	AFMC	Edwards AFB	F-16
85th TES	RegAF	ACC	Eglin AFB	Multiple
58th FS	RegAF	AETC	Eglin AFB	F-35A
40th FTS	RegAF	AFMC	Eglin AFB	F-35A
18th AGRS	RegAF	PACAF	Eielson AFB	F-16
194th FS	ANG	ANG	Fresno ANGB, CA	F-15C
190th FS	ANG	ANG	Gowen Field, Boise, ID	A-10C
199th FS	ANG	ANG	Hickam AFB	F-22A
4th FS	RegAF	ACC	Hill AFB	F-35A
34th FS	RegAF	ACC	Hill AFB	F-35A
421st FS	RegAF	ACC	Hill AFB	F-35A
311th FS	RegAF	ACC	Holloman AFB	F-16
314th FS	RegAF	ACC	Holloman AFB	F-16
93rd FS	AFR	AFRC	Homestead ARB, FL	F-16
159th FS	ANG	ANG	Jacksonville ANGB, FL	F-15C
182nd FS	ANG	ANG	JBSA Lackland - Kelly Field Annex	F-16
175th FS	ANG	ANG	Joe Foss Field, Sioux Falls, SD	F-16
90th FS	RegAF	PACAF	Joint Base Elmendorf-Richardson	F-22A
525th FS	RegAF	PACAF	Joint Base Elmendorf-Richardson	F-22A
114th FS	ANG	ANG	Kingsley Field, Klamath Falls, OR	F-15C
27th FS	RegAF	ACC	Joint Base Langley-Eustis	F-22A
94th FS	RegAF	ACC	Joint Base Langley-Eustis	F-22A
309th FS	RegAF	AETC	Luke AFB	F-16

Squadron	Component	Command	Current Base	MDS
310th FS	RegAF	AETC	Luke AFB	F-16
61st FS	RegAF	AETC	Luke AFB	F-35A
62nd FS	RegAF	AETC	Luke AFB	F-35A
157th FS	ANG	ANG	McEntire JNGB, SC	F-16
75th FS	RegAF	ACC	Moody AFB	A-10C
389th FS	RegAF	ACC	Mountain Home AFB	F-15E
391st FS	RegAF	ACC	Mountain Home AFB	F-15E
122nd FS	ANG	ANG	NAS JRB New Orleans, LA	F-15C
64th AGRS	RegAF	ACC	Nellis AFB	F-16
123rd FS	ANG	ANG	Portland, Oregon ANGB, OR	F-15C
107th FS	ANG	ANG	Selfridge ANGB, MI	A-10C
333rd FS	RegAF	ACC	Seymour Johnson AFB	F-15E
334rd FS	RegAF	ACC	Seymour Johnson AFB	F-15E
335th FS	RegAF	ACC	Seymour Johnson AFB	F-15E
336th FS	RegAF	ACC	Seymour Johnson AFB	F-15E
55th FS	RegAF	ACC	Shaw AFB	F-16
77th FS	RegAF	ACC	Shaw AFB	F-16
79th FS	RegAF	ACC	Shaw AFB	F-16
112th FS	ANG	ANG	Toledo ANGB, OH	F-16
148th FS	ANG	ANG	Tucson International Airport	F-16
152nd FS	ANG	ANG	Tucson International Airport	F-16
195th FS	ANG	ANG	Tucson International Airport	F-16
125th FS	ANG	ANG	Tulsa ANGB, Tulsa IA, OK	F-16
43rd FS	RegAF	ACC	Joint Base Langley-Eustis	F-22A
303rd FS	AFR	AFRC	Whiteman AFB	A-10C
63rd FS	RegAF	AETC	Luke AFB	F-35A
TBD	RegAF	AETC	Luke AFB	F-35A
TBD	RegAF	AETC	Luke AFB	F-35A
TBD	RegAF	AETC	Luke AFB	F-35A
31st FS	ANG	ANG	Montgomery ANGB, Montgomery Regional Airport, AL	F-35A
21st FS	RegAF	PACAF	Eielson AFB	F-35A
22nd FS	RegAF	PACAF	Eielson AFB	F-35A
TBD	AFR	AFRC	NAS JRB Fort Worth, TX[a]	F-35A
TBD	ANG	ANG	Truax Field ANGB, Madison, WI	F-35A
162nd FS	ANG	ANG	Tucson International Airport	F-16
422nd TES	RegAF	ACC	Nellis AFB	Multiple
57th WG	RegAF	ACC	Nellis AFB	Multiple
461st FTS	RegAF	AFMC	Edwards AFB	F-35A

Squadron	Component	Command	Current Base	MDS
163rd FS	ANG	ANG	Fort Wayne Intnl, Fort Wayne, IN	F-16
TBD	RegAF	ACC	Tyndall AFB	F-35A
TBD	RegAF	ACC	Tyndall AFB	F-35A
TBD	RegAF	ACC	Tyndall AFB	F-35A

NOTES: TBD = to be determined.
[a] As of the time of this report, the location of the USAFR F-35 was yet to be determined, but we chose Fort Worth in this analysis for illustrative purposes.

Appendix B. GCMs and Risk Metrics

Table B.1. List of GCMs and Source

GCM Name	Source
bcc-csm1-1	Beijing Climate Center, China Meteorological Administration, China
CanESM2	National Center for Atmospheric Research (NCAR), United States
CNRM-CM5	CERFACS (Centre National de Recherches Météorologiques, France)
CSIRO-Mk3-6-0	Australian Commonwealth Scientific and Industrial Research Organization
GFDL-ESM2M	Geophysical Fluid Dynamics Laboratory, United States
HadGEM2-CC365	Met Office Hadley Center, United Kingdom
HadGEM2-ES365	Met Office Hadley Center, United Kingdom
inmcm4	Institute for Numerical Mathematics, Russia
IPSL-CM5A-MR	Institut Pierre Simon Laplace, France
IPSL-CM5B-LR	Institut Pierre Simon Laplace, France
MIROC5	Atmosphere and Ocean Research Institute (University of Tokyo), National Institute for Environmental Studies, and Japan Agency for Marine-Earth Science and Technology
MIROC-ESM-CHEM	Japan Agency for Marine-Earth Science and Technology, Atmosphere and Ocean Research Institute (University of Tokyo), and National Institute for Environmental Studies

Table B.2. Power Reliability Metric Data for USAF Bases and Ranges

USAF Installation Type	Name	Average SAIFI (number of disruptions)	Average SAIDI (minutes)	Average Weighted FOR
Base	Atlantic City International Airport	1.49	403.32	13.19
Base	Barnes ANGB	0.93	119.14	11.56
Base	Buckley AFB	1.08	112.85	10.71
Base	Burlington ANGB	2.72	1,744.41	7.84
Base	Davis-Monthan AFB	0.72	71.65	10.06
Base	DC National Guard, JB Andrews	0.73	100.99	17.70
Base	Duluth ANGB, MN	1.53	503.62	7.54
Base	Edwards AFB	1.02	120.81	10.26
Base	Eglin AFB	1.23	114.74	8.62
Base	Eielson AFB	3.07	644.22	8.55
Base	Fort Wayne Intnl, Fort Wayne, IN	1.56	316.82	9.19
Base	Fresno ANGB, CA	1.55	374.20	6.20
Base	Gowen Field, Boise, ID	1.37	186.63	8.61

USAF Installation Type	Name	Average SAIFI (number of disruptions)	Average SAIDI (minutes)	Average Weighted FOR
Base	Hickam AFB	1.54	132.02	11.77
Base	Hill AFB	1.55	247.75	7.83
Base	Holloman AFB	1.21	91.32	11.70
Base	Homestead ARB, FL	1.33	915.46	16.84
Base	Jacksonville ANGB, FL	3.54	218.15	11.35
Base	JBSA Lackland - Kelly Field Annex	1.41	153.23	12.66
Base	JBSA Randolph	1.41	153.23	12.66
Base	Joe Foss Field, Sioux Falls, SD	2.08	243.87	7.89
Base	Joint Base Elmendorf-Richardson	1.80	235.52	16.86
Base	Kingsley Field, Klamath Falls, OR	1.55	247.75	7.83
Base	Joint Base Langley-Eustis	1.39	199.14	9.86
Base	Luke AFB	0.89	108.70	7.60
Base	McEntire JNGB, SC	2.55	925.55	9.74
Base	Montgomery ANGB, Montgomery Regional Airport, AL	1.85	222.49	8.02
Base	Moody AFB	1.72	525.27	7.85
Base	Mountain Home AFB	1.37	186.63	8.61
Base	NAS JRB Ft. Worth, TX	1.74	242.62	16.86
Base	NAS JRB New Orleans, LA	1.98	387.77	12.05
Base	Nellis AFB	0.49	51.28	13.65
Base	Portland, Oregon ANGB, OR	0.88	228.80	9.32
Base	Selfridge ANGB, MI	1.14	590.80	7.81
Base	Seymour Johnson AFB	1.84	603.00	8.34
Base	Shaw AFB	2.55	925.55	8.34
Base	Sheppard AFB	1.74	242.62	16.86
Base	Toledo ANGB, OH	0.78	122.07	2.10
Base	Truax Field ANGB, Madison, WI	5.84	172.35	16.76
Base	Tucson International Airport	0.72	71.65	10.06
Base	Tulsa ANGB, Tulsa IA, OK	2.35	428.00	14.77
Base	Tyndall AFB	1.44	231.50	12.75
Base	Warfield ANGB, Middle River, MD	0.94	124.80	6.43
Base	Whiteman AFB	1.37	298.75	8.46
Range	Adirondack Range*	1.16	229.95	13.43
Range	AF Dare County Range*	3.88	1,804.51	16.69
Range	Airburst Range	4.78	414.44	2.79
Range	Avon Park Range	2.19	803.27	12.75
Range	Barry M. Goldwater Range*	0.89	108.70	7.60
Range	Belle Fourche ESS Range*	1.08	88.31	10.77

USAF Installation Type	Name	Average SAIFI (number of disruptions)	Average SAIDI (minutes)	Average Weighted FOR
Range	Bollen Range	1.37	228.53	14.31
Range	Cannon Range	1.74	285.39	16.86
Range	Claiborne Range*	1.98	387.77	12.05
Range	Eastern Launch Facility	1.88	915.46	13.30
Range	Edwards Range	1.02	120.81	10.26
Range	Eglin Test and Training Complex*	1.23	114.74	8.62
Range	Falcon Range	2.82	455.52	14.77
Range	Grand Bay Range*	1.72	525.27	7.85
Range	Grayling Range	1.90	563.22	13.55
Range	Hardwood Range (Volk Field)*	1.46	277.90	11.13
Range	Holloman Range (aggregate)	1.21	91.32	11.70
Range	Indiana Air Range Complex - Atterbury	2.58	367.21	9.19
Range	Jefferson Range	2.34	407.40	9.19
Range	Joint Pacific Alaska Range Complex (aggregate)*	3.07	644.22	8.55
Range	McMullen Range (aggregate)	2.79	870.41	12.99
Range	Melrose Range*	1.28	186.33	10.56
Range	Mountain Home Range Complex (aggregate)*	1.37	186.63	8.61
Range	Navy Dare County Range	3.88	1,804.51	16.69
Range	Nevada Test and Training Range*	0.49	51.28	13.65
Range	Poinsett Range*	2.55	925.55	8.34
Range	Razorback Range	2.90	463.20	12.99
Range	Shelby Range	2.26	260.01	12.52
Range	Smoky Hill Range*	2.16	436.72	16.67
Range	Snyder Electronic Warfare Site*	3.61	807.54	8.61
Range	Utah Test and Training Range (aggregate)*	1.55	247.75	7.83
Range	Warren Grove Range*	1.49	403.32	13.19

* = Range is prioritized for upgrade by AF/A3T.

Table B.3. Hazard Exposure for USAF Air Bases and Ranges

USAF Installation Type	Name	Located in 3-ft SLR Inundation Zone?	Located in 7-ft SLR Inundation Zone?	Located in 10-ft SLR Inundation Zone?	Wildfire Potential	Seismic Risk Category[a]
Base	Atlantic City International Airport	No	No	No	Low to moderate	2
Base	Barnes ANGB	No	No	No	Nonburnable to very low	2
Base	Buckley AFB	No	No	No	Nonburnable to very low	2
Base	Burlington ANGB	No	No	No	Nonburnable to very low	3
Base	Davis-Monthan AFB	No	No	No	Nonburnable to low	3
Base	DC National Guard, JB Andrews	No	No	No	Nonburnable to very low	2
Base	Duluth ANGB, MN	No	No	No	Nonburnable to low	1
Base	Edwards AFB	No	No	No	Very low to low	15
Base	Eglin AFB	Yes	Yes	Yes	Very low to low	3
Base	Eielson AFB	No	No	No		
Base	Fort Wayne Intnl, Fort Wayne, IN	No	No	No	Nonburnable to very low	3
Base	Fresno ANGB, CA	No	No	No	Moderate	9
Base	Gowen Field, Boise, ID	No	No	No	High to very high	5
Base	Hickam AFB	Yes	Yes	Yes		
Base	Hill AFB	No	No	No	Very low to low	15
Base	Holloman AFB	No	No	No	Very low	3
Base	Homestead ARB, FL	Yes	Yes	Yes	Nonburnable to low	1
Base	Jacksonville ANGB, FL	No	No	No	Very low to high	3
Base	JBSA Lackland - Kelly Field Annex	No	No	No	Nonburnable to low	1
Base	JBSA Randolph	No	No	No	Very low to low	1
Base	Joe Foss Field, Sioux Falls, SD	No	No	No	Nonburnable to very low	2
Base	Joint Base Elmendorf-Richardson	No	No	No		
Base	Kingsley Field, Klamath Falls, OR	No	No	No	Nonburnable to very low	10
Base	Joint Base Langley-Eustis	Yes	Yes	Yes	Nonburnable to very low	2
Base	Luke AFB	No	No	No	Nonburnable to Moderate	3

USAF Installation Type	Name	Located in 3-ft SLR Inundation Zone?	Located in 7-ft SLR Inundation Zone?	Located in 10-ft SLR Inundation Zone?	Wildfire Potential	Seismic Risk Category[a]
Base	McEntire JNGB, SC	No	No	No	Moderate to high	3
Base	Montgomery ANGB, Montgomery Regional Airport, AL	No	No	No	Nonburnable to low	3
Base	Moody AFB	No	No	No	Nonburnable to moderate	3
Base	Mountain Home AFB	No	No	No	High	4
Base	NAS JRB Fort Worth, TX	No	No	No	Nonburnable to very low	2
Base	NAS JRB New Orleans, LA	Yes	Yes	Yes	Nonburnable to very low	2
Base	Nellis AFB	No	No	No	Very low to low	7
Base	Portland, OR ANGB, OR	Yes	Yes	Yes	Nonburnable to very low	10
Base	Selfridge ANGB, MI	No	No	No	Nonburnable to very low	2
Base	Seymour Johnson AFB	No	No	No	Nonburnable to very low	2
Base	Shaw AFB	No	No	No	Nonburnable to moderate	3
Base	Sheppard AFB	No	No	No	Low	3
Base	Toledo ANGB, OH	No	No	No	Nonburnable to very low	3
Base	Truax Field ANGB, Madison, WI	No	No	No	Nonburnable to very low	2
Base	Tucson International Airport	No	No	No	Nonburnable to low	3
Base	Tulsa ANGB, Tulsa IA, OK	No	No	No	Nonburnable to very low	3
Base	Tyndall AFB	Yes	Yes	Yes	Very low to moderate	2
Base	Warfield ANGB, Middle River, MD	Yes	Yes	Yes	Nonburnable to very low	2
Base	Whiteman AFB	No	No	No	Nonburnable to very low	3
Range	Adirondack Range	No	No	No	Very low to moderate	3
Range	AF Dare County Range	Yes	Yes	Yes	Very high	2
Range	Airburst Range	No	No	No	Very low to low	3
Range	Avon Park Range	No	No	No	High to very high	1
Range	Barry M. Goldwater Range (East)	No	No	No	Very low to low	5

USAF Installation Type	Name	Located in 3-ft SLR Inundation Zone?	Located in 7-ft SLR Inundation Zone?	Located in 10-ft SLR Inundation Zone?	Wildfire Potential	Seismic Risk Category[a]
Range	Belle Fourche ESS Range	No	No	No	Very low	2
Range	Bollen Range	No	No	No	Very low	2
Range	Cannon Range	No	No	No	Very low to low	4
Range	Claiborne Range	No	No	No	Low to high	3
Range	Eastern Launch Facility	Yes	Yes	Yes	Nonburnable to very low	1
Range	Edwards Range	No	No	No	Very low to low	15
Range	Eglin Test and Training Complex	No	Yes	Yes	Very low to moderate	3
Range	Falcon Range	No	No	No	Low to moderate	3
Range	Grand Bay Range	No	No	No	Low to high	3
Range	Grayling Range	No	No	No	Very low to moderate	2
Range	Hardwood Range (Volk Field)	No	No	No	Very low to moderate	2
Range	Holloman Range (aggregate)	No	No	No	Very low to low	3
Range	Indiana Air Range Complex - Atterbury	No	No	No	Very low	4
Range	Jefferson Range	No	No	No	Very low	4
Range	Joint Pacific Alaska Range Complex	No	No	No		
Range	McMullen Range (aggregate)	No	No	No	Very low to low	1
Range	Melrose Range	No	No	No	Moderate to high	2
Range	Mountain Home Range Complex (aggregate)	No	No	No	High to very high	4
Range	Navy Dare County Range	Yes	Yes	Yes	Moderate to very high	2
Range	Nevada Test and Training Range	No	No	No	Very low to moderate	9
Range	Poinsett Range	No	No	No	Low to high	4
Range	Razorback Range	No	No	No	Very low to moderate	4
Range	Shelby Range	No	No	No	Moderate to very high	3
Range	Smoky Hill Range	No	No	No	Moderate to high	2
Range	Snyder Electronic Warfare Site	No	No	No	Low to moderate	1
Range	Utah Test and Training Range (aggregate)	No	No	No	Nonburnable	4

USAF Installation Type	Name	Located in 3-ft SLR Inundation Zone?	Located in 7-ft SLR Inundation Zone?	Located in 10-ft SLR Inundation Zone?	Wildfire Potential	Seismic Risk Category[a]
Range	Warren Grove Range	No	No	No	Moderate to very high	2

SOURCES: Authors' analysis of NOAA SLR inundation maps from NOAA map server files, seismic hazard maps from USGS map server files, and wildfire potential hazard maps from USFS map server files (NOAA, Office for Coastal Management, undated-a; NOAA, Office for Coastal Management, undated-b; NOAA, Office for Coastal Management, undated-c; U.S. Department of Agriculture, U.S. Forest Service, undated; USGS National Seismic Hazard Mapping Program, undated).

[a] The seismic risk category is a measure of the peak acceleration expressed as a percent of gravity. The peak acceleration is the maximum acceleration that a particular location experiences during an earthquake, which is measured in centimeters per second squared.

Table B.4. Sample Downscaled Projections of Annual Cross-GCM Average Number of Black Flag Training Days in the Near Term

USAF Installation Type	Name	RCP 4.5	RCP 8.5
Base	Atlantic City International Airport	39.9	43.9
Base	Barnes ANGB	26.7	32.3
Base	Buckley AFB	7.1	11.0
Base	Burlington ANGB	11.8	15.8
Base	Davis-Monthan AFB	86.1	90.5
Base	DC National Guard, JB Andrews	53.1	58.1
Base	Duluth ANGB, MN	4.3	5.5
Base	Edwards AFB	61.4	69.4
Base	Eglin AFB	107.4	112.4
Base	Eielson AFB	No data	No data
Base	Fort Wayne Intnl, Fort Wayne, IN	35.4	40.4
Base	Fresno ANGB, CA	93.0	100.1
Base	Gowen Field, Boise, ID	22.1	27.2
Base	Hickam AFB	No data	No data
Base	Hill AFB	3.3	4.8
Base	Holloman AFB	67.8	70.3
Base	Homestead ARB, FL	144.3	151.1
Base	Jacksonville ANGB, FL	117.8	121.6
Base	JBSA Lackland - Kelly Field Annex	146.2	151.1
Base	JBSA Randolph	148.1	153.0
Base	Joe Foss Field, Sioux Falls, SD	36.0	39.6
Base	Joint Base Elmendorf-Richardson	No data	No data
Base	Kingsley Field, Klamath Falls, OR	9.0	13.0
Base	Joint Base Langley-Eustis	52.9	57.8

USAF Installation Type	Name	RCP 4.5	RCP 8.5
Base	Luke AFB	119.2	122.9
Base	McEntire JNGB, SC	101.8	106.7
Base	Montgomery ANGB, Montgomery Regional Airport, AL	115.4	119.0
Base	Moody AFB	125.9	129.5
Base	Mountain Home AFB	22.1	27.2
Base	NAS JRB Fort Worth, TX	118.0	121.6
Base	NAS JRB New Orleans, LA	124.2	127.4
Base	Nellis AFB	50.5	55.4
Base	Portland, Oregon ANGB, OR	6.9	9.8
Base	Selfridge ANGB, MI	23.2	27.6
Base	Seymour Johnson AFB	85.3	89.8
Base	Shaw AFB	92.8	97.9
Base	Sheppard AFB	115.1	117.4
Base	Toledo ANGB, OH	32.5	37.4
Base	Truax Field ANGB, Madison, WI	26.5	30.8
Base	Tucson International Airport	91.8	95.9
Base	Tulsa ANGB, Tulsa IA, OK	93.9	95.2
Base	Tyndall AFB	101.5	105.7
Base	Warfield ANGB, Middle River, MD	54.7	59.8
Base	Whiteman AFB	67.3	69.6
Range	Adirondack Range	10.3	13.5
Range	AF Dare County Range	73.7	79.0
Range	Airburst Range	19.4	25.3
Range	Avon Park Range	163.6	168.4
Range	Barry M. Goldwater Range (East)	132.5	136.3
Range	Belle Fourche ESS Range	32.5	37.0
Range	Bollen Range	24.3	29.6
Range	Cannon Range	71.6	73.7
Range	Claiborne Range	130.0	132.5
Range	Eastern Launch Facility	97.5	106.2
Range	Edwards Range	114.3	118.4
Range	Eglin Test and Training Complex	102.2	104.0
Range	Falcon Range	127.2	130.8
Range	Grand Bay Range	14.3	18.1
Range	Grayling Range	25.8	30.4
Range	Hardwood Range (Volk Field)		
Range	Holloman Range (aggregate)	71.0	73.6

USAF Installation Type	Name	RCP 4.5	RCP 8.5
Range	Indiana Air Range Complex - Atterbury	52.1	56.5
Range	Jefferson Range	53.3	58.0
Range	Joint Pacific Alaska Range Complex (aggregate)	No data	No data
Range	McMullen Range (aggregate)	173.7	178.1
Range	Melrose Range	73.9	77.0
Range	Mountain Home Range Complex (aggregate)	27.8	33.9
Range	Navy Dare County Range	75.6	80.6
Range	Nevada Test and Training Range	0.0	0.1
Range	Poinsett Range	104.8	109.7
Range	Razorback Range	104.1	106.7
Range	Shelby Range	131.9	135.1
Range	Smoky Hill Range	79.5	81.9
Range	Snyder Electronic Warfare Site	107.5	111.1
Range	Utah Test and Training Range (aggregate)	13.7	18.5
Range	Warren Grove Range	46.3	50.3

References

AF/A3T—*See* Office of the Director of Training and Readiness, Deputy Chief of Staff for Operations, Headquarters U.S. Air Force.

AFI—*See* Air Force Instruction.

AFMAN—*See* Air Force Manual.

Air Combat Command, Flight Operations Division, *F-22A Ready Aircrew Program (RAP) Tasking Memorandum, Aviation Schedule 2019 (AS-19)*, memorandum for F-22 Operations Group Commanders, October 2, 2018.

Air Force Instruction 10-201, *Force Readiness Reporting*, January 8, 2019.

Air Force Instruction 10-503, *Strategic Basing*, July 28, 2017.

Air Force Instruction 11-2F-16, *F-16 Pilot Training*, Vol. 1, Air Combat Command, Flight Operations and Training Branch, April 20, 2015.

Air Force Instruction 16-401, *Designating and Naming Defense Military Aerospace Vehicles*, May 16, 2014.

Air Force Instruction 48-151, *Thermal Injury Prevention Program*, April 2016.

Air Force Manual 13-212, *Range Planning and Operations*, Vol. 1, June 22, 2018.

Air Force Manual 32-1084, *Civil Engineering: Facility Requirements*, February 26, 2016.

Ausink, John, A. Anthony D. Rosello, Timothy Marler, Michael Vasseur, Brynn Tannehill, Dara Gold, Kelly Klima, and Laura Kupe, *Fifth-Generation Aircraft Operational Training Infrastructure: Practices, Gaps, and Proposed Solutions*, Santa Monica, Calif.: RAND Corporation, 2018, Not available to the general public.

Ausink, John A., William W. Taylor, James H. Bigelow, and Kevin Brancato, *Investment Strategies for Improving Fifth-Generation Fighter Training*, Santa Monica, Calif.: RAND Corporation, TR-871-AF, 2011. As of August 18, 2020: https://www.rand.org/pubs/technical_reports/TR871.html

Bednarz, Sean, Anu Narayanan, Paul DeLuca, Joshua Baron, Robert A. Guffey, Daniel M. Romano, and Joseph V. Vesely, *Shaping Domestic Basing Decisions to Ensure Affordable Readiness: A Screening-Level Assessment of Alternative Domestic Basing Postures for the F-35A*, Santa Monica, Calif.: RAND Corporation, 2016, Not available to the general public.

Department of Defense Instruction 4165.03, *DoD Real Property Categorization*, Washington, D.C.: U.S. Department of Defense, February 4, 2015.

DoD—*See* U.S. Department of Defense.

DoDI—*See* Department of Defense Instruction.

Gould, Joe, "Storm-Ravaged Bases Wait on Washington for Repair Money," *Defense News*, May 10, 2019.

Hajizadeh, Roohalah, Somayeh Farhang Dehghan, Farideh Golbabaei, Sayed Mohammad Jafari, and Mehrdad Karajizadeh, "Offering a Model for Estimating Black Globe Temperature According to Meteorological Measurements," *Meteorological Applications*, Vol. 24, 2017, pp. 303–307.

Hannah, Lee, *Climate Change Biology*, 2nd ed., London: Academic Press, 2015. As of October 5, 2019: https://www.sciencedirect.com/book/9780124202184/climate-change-biology

Lanzante, John R., Keith W. Dixon, Mary Jo Nath, Carolyn E. Whitlock, and Dennis Adams-Smith, "Some Pitfalls in Statistical Downscaling of Future Climate," *Bulletin of the American Meteorological Society*, Vol. 99, No. 4, April 2018, pp. 791–803. As of October 5, 2019: https://doi.org/10.1175/BAMS-D-17-0046.1

Lostumbo, Michael J., Michael J. McNerney, Eric Peltz, Derek Eaton, David R. Frelinger, Victoria A. Greenfield, John Halliday, Patrick Mills, Bruce R. Nardulli, Stacie L. Pettyjohn, Jerry M. Sollinger, and Stephen M. Worman, *Overseas Basing of U.S. Military Forces: An Assessment of Relative Costs and Strategic Benefits*, Santa Monica, Calif.: RAND Corporation, RR-201-OSD, 2013. As of August 31, 2020: http://www.rand.org/pubs/research_reports/RR201.html

McGarvey, Ronald G., James H. Bigelow, Gary James Briggs, Peter Buryk, Raymond E. Conley, John G. Drew, Perry Shameem Firoz, Julie Kim, Lance Menthe, S. Craig Moore, William W. Taylor, and William A. Williams, *Assessment of Beddown Alternatives for the F-35*, Santa Monica, Calif.: RAND Corporation, RR-124-AF, 2013. As of August 31, 2020: https://www.rand.org/pubs/research_reports/RR124.html

Mills, Patrick, Adam Grissom, Jennifer Kavanagh, Leila Mahnad, and Stephen M. Worman, *A Cost Analysis of the U.S. Air Force Overseas Posture: Informing Strategic Choices*, Santa Monica, Calif.: RAND Corporation, RR-150-AF, 2013. As of August 18, 2020: https://www.rand.org/pubs/research_reports/RR150.html

Narayanan, Anu, Sean Bednarz, John A. Ausink, Joshua Baron, Anthony DeCicco, Robert A. Guffey, George E. Hart, John Matsumura, Michael Nixon, Chuck Stelzner, William W. Taylor, and Joseph V. Vesely, *A Modeling Framework for Optimizing F-35A Strategic Basing Decisions to Meet Training Requirements*, Santa Monica, Calif.: RAND Corporation,

RR-1546-AF, 2016. As of August 18, 2020:
https://www.rand.org/pubs/research_reports/RR1546.html

Narayanan, Anu, Debra Knopman, James D. Powers, Bryan Boling, Benjamin M. Miller, Patrick Mills, Kristin Van Abel, Katherine Anania, Blake Cignarella, and Connor P. Jackson, *Air Force Installation Energy Assurance: An Assessment Framework*, Santa Monica, Calif.: RAND Corporation, RR-2066-AF, 2017. As of August 18, 2020:
https://www.rand.org/pubs/research_reports/RR2066.html

National Oceanic and Atmospheric Administration, Office for Coastal Management, "dc_slr/slr_3ft (MapServer)," webpage, undated-a. As of August 31, 2020:
https://coast.noaa.gov/arcgis/rest/services/dc_slr/slr_3ft/MapServer

———, "dc_slr/slr_7ft (MapServer)," webpage, undated-b. As of August 31, 2020:
https://coast.noaa.gov/arcgis/rest/services/dc_slr/slr_7ft/MapServer

———, "dc_slr/slr_10ft (MapServer)," webpage, undated-c. As of August 31, 2020:
https://coast.noaa.gov/arcgis/rest/services/dc_slr/slr_10ft/MapServer

NOAAClimate.gov, "Climate Models," webpage, undated. As of August 31, 2020:
https://www.climate.gov/maps-data/primer/climate-models

NOAA, Office for Coastal Management—*See* National Oceanic and Atmospheric Administration, Office for Coastal Management.

North American Electric Reliability Corporation, "Generating Availability Data System (GADS)," webpage, undated. As of August 31, 2020:
https://www.nerc.com/pa/RAPA/gads/Pages/GeneratingAvailabilityDataSystem-(GADS).aspx

Northwest Knowledge Center, "Multivariate Adaptive Constructed Analogs (MACA): Dataset," webpage, undated. As of October 5, 2019:
https://climate.northwestknowledge.net/MACA/

Office of the Director of Training and Readiness, Deputy Chief of Staff for Operations, Headquarters U.S. Air Force, "Range Plan v15 11 Feb 19," spreadsheet, February 11, 2019, Not available to the general public.

Resource Adequacy Planning, *PJM Manual 22: Generator Resource Performance Indices*, Rev. 18, Audobon, Pa.: PJM, 2017. As of October 5, 2019:
https://www.pjm.com/~/media/documents/manuals/m22.ashx

Rosello, Anthony D., Bradley DeBlois, Michael McGee, James Williams, Chad J. R. Ohlandt, Muharrem Mane, Bart E. Bennett, C. R. Anderegg, Andrew Cady, Jeffrey Kendall, Robert A. Guffey, and Michael Bohnert, *Adversary Air Enterprise: Cost-Effective Options for the U.S.*

Air Force, Santa Monica, Calif: RAND Corporation, 2019, Not available to the general public.

Samaras, Constantine, Rachel Costello, Paul DeLuca, Stephen J. Guerra, Kenneth Kuhn, Anu Narayanan, Michael Nixon, Stacie L. Pettyjohn, Nolan Sweeney, Joseph V. Vesely, and Lane F. Burgette, *Improvements to Air Force Strategic Basing Decisions*, Santa Monica, Calif.: RAND Corporation, RR-1297-AF, 2016. As of August 18, 2020: https://www.rand.org/pubs/research_reports/RR1297.html

Stull, Roland, "Wet-Bulb Temperature from Relative Humidity and Air Temperature," *Journal of Applied Meteorology and Climatology*, Vol. 50, No. 11, 2011, pp. 2267–2269. As of October 5, 2019: https://doi.org/10.1175/JAMC-D-11-0143.1

USAF—*See* U.S. Air Force.

U.S. Air Force, *Fiscal Year (FY) 2019 Budget Estimates, Military Personnel Appropriation*, February 2018.

U.S. Air Force, Directorate of Civil Engineers (AF/A4C), "Civil Engineering Planning Template," spreadsheet provided to authors, January 16, 2019.

U.S. Department of Agriculture, U.S. Forest Service, "USDA_USFS_2014_Wildfire_Hazard_Potential (MapServer)," webpage, undated. As of October 5, 2019: https://maps7.arcgisonline.com/arcgis/rest/services/USDA_USFS_2014_Wildfire_Hazard_Potential/MapServer

U.S. Department of Defense, *Base Closure and Realignment Report,* Volume I, *Part 1 of 2: Results and Process*, May 2005. As of September 10, 2019: https://web.archive.org/web/20050519163315/http://www.defenselink.mil/brac/pdf/Vol_I_Part_1_DOD_BRAC.pdf

———, *Unified Facilities Criteria 3-701-0123*, May 2018, change 1, June 25, 2018.

———, *Report on Effects of a Changing Climate to the Department of Defense*, Washington, D.C.: Office of the Under Secretary of Defense for Acquisition and Sustainment, January 2019.

U.S. Energy Information Administration, "Form EIA-861: Annual Electric Power Industry Report Instructions," 2017. As of October 5, 2019: https://elecidc12c.eia.doe.gov/2017 EIA-861 Instructions.pdf

———, "Average U.S. Electricity Customer Interruptions Totaled Nearly 8 Hours in 2017," *Today in Energy* blog, November 30, 2018. As of October 9, 2019: https://www.eia.gov/todayinenergy/detail.php?id=37652

U.S. Geological Survey National Seismic Hazard Mapping Program, "haz/US1hz050_2014 (MapServer)," webpage, undated. As of October 5, 2019: https://earthquake.usgs.gov/arcgis/rest/services/haz/US1hz050_2014/MapServer

USGS National Seismic Hazard Mapping Program—*See* U.S. Geological Survey National Seismic Hazard Mapping Program.